Multimodal Biometrics and Intelligent Image Processing for Security Systems

Marina L. Gavrilova
University of Calgary, Canada

Maruf Monwar
Carnegie Mellon University, USA

T0338712

Information Science
REFERENCE

Managing Director:	Lindsay Johnston
Editorial Director:	Joel Gamon
Book Production Manager:	Jennifer Yoder
Publishing Systems Analyst:	Adrienne Freeland
Development Editor:	Myla Merkel
Assistant Acquisitions Editor:	Kayla Wolfe
Typesetter:	Christina Henning
Cover Design:	Jason Mull

Published in the United States of America by
Information Science Reference (an imprint of IGI Global)
701 E. Chocolate Avenue
Hershey PA 17033
Tel: 717-533-8845
Fax: 717-533-8661
E-mail: cust@igi-global.com
Web site: http://www.igi-global.com

Library of Congress Cataloging-in-Publication Data

Gavrilova, Marina L.Multimodal biometrics and intelligent image processing for security systems / by Marina L. Gavrilova and Maruf Monwar.
pages cm
Includes bibliographical references and index.
Summary: "This book provides an in-depth description of existing and fresh fusion approaches for multimodal biometric systems, covering relevant topics affecting the security and intelligent industries"-- Provided by publisher.
ISBN 978-1-4666-3646-0 (hardcover) -- ISBN 978-1-4666-3647-7 (ebook) -- ISBN 978-1-4666-3648-4 (print & perpetual access) 1. Biometric identification. I. Monwar, Maruf, 1976- II. Title.
TK7882.B56G385 2013
006.4--dc23
2012048472

British Cataloguing in Publication Data
A Cataloguing in Publication record for this book is available from the British Library.

All work contributed to this book is new, previously-unpublished material. The views expressed in this book are those of the authors, but not necessarily of the publisher.

With all my heart, I dedicate this book to those people in my life who made my everyday learning worth learning: my parents Lev and late Tatiana Felman for their unconditional support, my grandparents, especially grandmother Alexandra Fedorovna Pestryakova for her amazing energy and keeping focus in life, my husband Dmitri Gavrilov for keeping me in focus, and my incredible sons, Andrei and Artemy, for being who they are.

I also would like to dedicate it to my teachers both at Lomonosov Moscow State University and University of Calgary for making learning a fun pastime, and my friends and colleagues in Russia, Canada, and around the globe for being unique and wonderful individuals.

Marina L. Gavrilova
University of Calgary, Canada

It gives me a great pleasure to dedicate this book to my entire family, especially my wife, Nahid Sultana, and my sweet little daughter, Rushama Nahiyan, for their incredible love, prayers, enthusiasm, and encouragement. Without their support, it would not have been possible for me.

Maruf Monwar
Carnegie University, USA

Table of Contents

Section 1
Biometric Overview and Trends

Chapter 3

Section 2
Current Practices in
Information Fusion for Multimodal Biometrics

Chapter 4

Chapter 5

Chapter 6

Section 3
Applications in Security Systems

Chapter 11

Foreword

Recently there has been growing interest in "Biometrics," which deals with the study of computer emulation, analysis, and synthesis of human behavior that is related to the human body. It has many significant applications in solving daily problems, including personal verification and identification, analysis and recognition of handwriting, signature, fingerprint, voice, speech, palm and iris, and cybersecurity. This is particularly important in national security, especially after the 911 tragedy at New York's World Trade Center. Like the famous doctors' saying, "It is more important to prevent illness beforehand than healing it afterwards," many believe that if we had good enough "biometrics" system to prevent those "terrorists" from entering the USA, there would have been no 911 tragedy.

I am pleased to see that there is a new book coming out in this field, by renowned Prof. Marina L. Gavrilova at University of Calgary, Canada, to be published by the IGI Global Publisher. I have met Prof. Gavrilova while I was visiting Biometric technology lab at the University of Calgary as iCORE (Informatics Circle of Research Excellence), and have been quite impressed by her work on "Emotion Analysis and Recognition."

This book covers a rather wide range of "Biometrics," both in fundamental theories and practical applications, including its overview, relation to Artificial Intelligence (AI) and Pattern Recognition (PR), Image Processing (IP), and Neuron Networks, Biometrics Systems, Current Practice in Information Fusion and Multimodal Biometrics, Trend, Fuzzy Fusion Biometrics, and finally Applications in Security Systems and Robotics.

I am particularly impressed by its skillfully designed AI technique-decision making system for biometrics verification and identification including "Face," Ear," and "Iris" multi-modal biometric information fusion system, using Markov Chain and Fuzzy logic. It makes the system more intelligent (smart) and can enhance recognition accuracy rate.

In its final chapter of conclusions and future directions, there are many exciting topics as future research for MS or PhD level graduate students.

Overall, I think this book is good for researchers and professionals who are interested in Biometrics, and for senior undergraduate and graduate students at both MS and PhD levels. It can serve as a good textbook as well as a research reference.

Patrick S.P. Wang
WANG Teknowloge Lab, USA & Northeastern University, USA & Harvard University, USA
& University of Calgary, Canada & Magdeburg University, Germany
Fellow, IAPR, ISIBM, WASE, IEEE, ISIBM Distinguished Achievement Awardee,
Founding EIC, IJPRAI and WSP Book Series on MPAI

Quotes and Testimonials

QUOTES

Knowledge is a tool, not the goal.

Lev Nikolaevich Tolstoi

Data is not information, information is not knowledge, knowledge is not understanding, understanding is not wisdom.

Clifford Stoll

Education is the most powerful weapon which you can use to change the world.

Nelson Mandela

TESTIMONIALS

This book is the result of comprehensive study done by the authors to understand and demonstrate curious relationships between computational intelligence and biometric security. It is a must have for a researcher looking for the most up to date advancements.

Prof. Alexei Sourin, NTU Singapore, Founder and Chair of *CyberWorlds Conference* Series

Artificial Intelligence techniques have been used more and more in various research areas during these last two decades. The use of these techniques permitted researchers to find smart and useful solutions for a lot of problems, especially in Computer Graphics and Image Processing. Several Artificial Intelligence techniques are nowadays used in these areas, such as neural networks, genetic algorithms, constraint satisfaction techniques, heuristic search, fuzzy logic, and many others.

Five years ago, I wrote a book on improvements of Computer Graphics techniques based on the use of various Artificial Intelligence methods, and I am very happy to see that Prof. Marina Gavrilova decided to write this book from the point of view of intelligent Image Processing. For several years, I have known and appreciated Prof. Gavrilova's research work, and I think that she is the most qualified person for writing such an important and useful book.

Prof. Dimitri Plemenos, University of Limoges, France, Creator of the 3IA *International Annual Conference on Intelligent Computer Graphics*

The book is a tremendous achievement and a marvelous manuscript about Biometrics. Not only have the authors explained in a comprehensible way what exactly Biometrics is, but they also have added many invaluable new aspects and tools to it.

Prof. Khalid Saeed, AHG Poland, Author of *Biometrics and Kansei Engineering*, Springer

I found this book to be exceptional for more than one reason: the applications addressed are timely; the book provides a mature coverage of multi-disciplinary topics illustrating the cross-fertilization across sub-disciplines; and it is quite self-contained. The book presents state-of-the-art methodologies to biometric security decision-making processes. This publication is a great candidate for graduate level courses as well as for educators, scholars, and practitioners. This is the most informative book I have seen in this field of study.

Prof. Hamid R. Arabnia, PhD., The University of Georgia Department of Computer Science, Editor-in-Chief of *The Journal of Supercomputing*, Springer, Co-Editor/Board of The Journal of Computational Science, Elsevier, Elected Fellow of the Int'l Society of Intelligent Biological Medicine (ISIBM)

Preface

It is hard to say what started first, the perpetual drive of humans to understand their place in the world, or the perpetual challenges the world gives to humans, forcing us to evolve as intelligent beings capable of critically analyzing our environment and making (for the most part) sensible decisions. While this question might never be fully answered by historians and philosophers, it provides a rich soil on which fruits of knowledge could be pastured, cherished, gathered, and shared freely among others.

Over the course of human history, sharing such knowledge has taken on many different forms: from pre-historic cave drawings found in Andalusia, Spain, to Egyptian clay tables now on display in Louvre, France, from Ancient Mediterranean papyrus writing to encyclopedia Britannica, from Julius Ceasar's Roman calendar to Wikipedia, there are various forms the knowledge can take. While some of its most noble carriers who rightfully take their place in history believe in sharing it freely with the world—through pupil tutoring, creation of academies, and free public lectures—some also realize the power and competitive edge the knowledge can bring. Thus, almost as old as writing, encryption methods have been introduced; secret files and protocols were established, and some literary works were prohibited from being displayed in public. Ceasar's ciphers, Enigma machines, RSA private-public encryption, digital signatures, watermarking, and firewalls are only a few examples of how information can be protected from unwanted eyes. Thus, the area of biometrics comes to light, addressing the unresolved balance between the privacy and security, between overt and covert data collection, between high secrecy surrounding some algorithm developments, and the need to disseminate the research to the wildest population possible.

The foundation for this book is the research conducted at the Biometric Technologies Laboratory, established at the University of Calgary in 2001. The methodology presented in Part I and Part III of the book introduces reader to the biometric field and describes general directions of research conducted by members of BT lab. It includes applications of neural networks, intelligent processing, context-based methods to security problems in biometric recognition and virtual reality domains. Part II of the book is written in collaboration with a doctoral student of the BT Lab, working on Markov Chains and Fuzzy logic for rank-level fusion.

It gives me great pleasure to invite the reader for a journey into the area of biometric security, intelligent agents, self-replicating robots, virtual entities, multi-modal biometrics, data fusion, fuzzy logic, neural networks, and social behavioral profiling, and leaving him/her to emerge with perhaps a slightly changed perception of the world and his/her own journey in it.

Marina L. Gavrilova
University of Calgary, Canada
November 14, 2012

Acknowledgment

The first author would like to acknowledge the contributions of all the members of Biometric Technologies Laboratory (BTLab) at the University of Calgary. We express deep appreciation to lab member Ph.D. Kushan Ahmadian for his contribution to development of neural network methodology chapter. We also would like to thank lab member M.Sc. student Shermin Bazazian for gait analysis method implementation.

We would like to deeply thank Prof. Yingxu Wang, Electrical Engineering Department, University of Calgary, for his insightful suggestions and collaboration on multimodal fuzzy system. We would like to express our gratitude to Anil Jain and Roman Yampolskii for their passion for biometrics and artimetrics. Huge thanks go to Svetlana Yanushkevich and Vladimir Shmerko for starting it all with embracing biometric subject, and also for Patrick Wang and Sargur Srihari for their contribution to the Inaugural Biometric Workshop and the subsequent book on biometric synthesis. We also acknowledge all other wonderful individuals, including Khalid Saeed (Poland), Hamid Arabnia (USA), Alexey Sourin (Singapore), Dimitri Plemenos (France), Christos Papadimitriou (USA) for their profound knowledge and never ending enthusiasm.

Funding agencies supporting this project were Canadian Foundation for Innovation, NSERC, NATO, MITACS, Alberta Ingenuity, and PIMS.

In addition, the second author would like to express his deepest gratitude to all of his teachers at the University of Rajshahi, Bangladesh; University of Northern BC, Canada; and University of Calgary, Canada, especially to Prof. Vijayakumar Bhagavatula, MSU; Prof. Marina L. Gavrilova, Prof. Yingxu Wang, Prof. Jon Rokne, and Dr. Steve Liang, University of Calgary; and Prof. Piotr Porwik, University of Silesia, Katowice, Poland for their guidance and excellent support rendered over the past several years. He is thankful to the Natural Science and Engineering Research Council of Canada (NSERC, Canada) and Innovates Centre for Research Excellence (iCORE), Alberta, Canada, for partially supporting his research.

We thank anonymous book reviewers for their valuable comments and all IGI community members: staff, publishers, editors, typesetters, communicators, for their patience and unlimited support of this project.

To all of them, we express our sincere gratitude. Last but not the least, immense appreciation goes to our families for their support; without them, this endeavor would have never been possible.

Marina L. Gavrilova
University of Calgary, Canada

Section 1
Biometric Overview and Trends

Chapter 1
Introduction

ABSTRACT

This chapter presents an introductory overview of the application of computational intelligence in biometrics. Starting with the historical background on artificial intelligence, the chapter proceeds to the evolutionary computing and neural networks. Evolutionary computing is an ability of a computer system to learn and evolve over time in a manner similar to humans. The chapter discusses swarm intelligence, which is an example of evolutionary computing, as well as chaotic neural network, which is another aspect of intelligent computing. At the end, special concentration is given to a particular application of computational intelligence—biometric security.

1. A HISTORICAL LOOK AT ARTIFICIAL INTELLIGENCE

Over the course of the human history, the greatest minds: scientists, philanthropists, educators, politicians, leaders, philosophers, were fascinated with the way human brain works. From Michelangelo to Lomonosov, from DaVinci to Einstein, there have been numerous attempts to uncover the mystery of human mind and to replicate its working first through simple mechanical devices and later, in the 20th century, through computing machines and intelligent software.

In Alan Turing's groundbreaking work "Computing Machinery and Intelligence," he posed the question "Can machines think?" (Turing, 1950). In order to establish credible criteria to answer this question, he proposed a test, now well known as "The Turing Test" – to estimate a machine's ability to demonstrate intelligence. The test is based on a conversation in a natural language between the human judge and the opponent, who can be either human or a machine. Judging by the answers, the judge must distinguish the machine from the human. If the judge fails to do so, the machine is deemed to have passed the test.

DOI: 10.4018/978-1-4666-3646-0.ch001

After the theoretical platform for an Automated Turing Test (ATT) was developed by Naor in 1996 (Naor, 1996), the new generation of researchers continued to study the same concept of human/machine identification. In addition to ATT, the new developed procedures were "Reversed Turing Test" (RTT); "Human Interactive Proof" (HIP); "Mandatory Human Participation" (MHP); and the "Completely Automated Public Turing Test to tell Computers and Humans Apart" (CAPTCHA) (Ahn, Blum, Hopper, & Langford, 2003).

In the modern terms, the Turing test could be considered as one of the behavioral biometrics, whereas the behavior is based on the literary responses to the questions. Another groundbreaking work in *modern artificial intelligence* was laid out by John von Neumann in the 1950's in his theory of automata and self-replicating machines, later published as a book (von Neumann, 1966). His theoretical concepts were based on those of Alan Turing.

Self-replication is a natural process which lies at the base of biological life cycle on Earth. At the core of self-replication of living organisms lies the biological fact that nucleic acids produce copies of themselves under proper conditions (Craig, Cohen-Fix, Green, Greider, Storz, & Wolberger, 2011). Self-replication in non-biological contexts has been studied only recently in the context of "artificial" entities such as self-replicating software, computer viruses, and robots (Gavrilova & Yampolskiy, 2011). The research has been fruitful in the past decade. Cornell University, Canada researchers have created a machine that can build copies of itself. Their robots are made up of a series of modular cubes (called "molecubes"), each containing identical parts and the computer program for replication. The cubes can change their topology by selectively attaching and detaching from each other using implanted magnets, where a complete robot is made up of a number of cubes connected together (Zykov, Mytilinaios, Adams, & Lipson, 2005).

2. EVOLUTIONAL COMPUTING AND NEURAL NETWORKS

Another approach taken in the same direction has led to a concept of *evolutionary computing*. The concept exploited here is ability of computer software to learn and evolve over time, similarly to how human learns, from experiences, facts and by example. The ability to develop winning strategies and improve itself comes as a natural, even so sometimes surprising result.

As an example, *Swarm Intelligence* (SI) is one of subareas of evolutionary computing, whereas the system functions based on the collective behaviors of (unsophisticated) agents interacting locally with their environment (Bonabeau, Dorigo, & Theraulaz, 1999). Agents in Swarm Intelligence system have limited perception (intelligence) and cannot individually carry out the task it intends to. Nevertheless, by regulating the behavior of agents in a swarm, one can demonstrate emergent behavior and intelligence as a collective phenomenon. Although the swarming phenomenon is largely observed among biological organisms such as an ant colony, a flock of birds etc., it is recently being used to simulate complex dynamic systems focused towards accomplishing a well-defined objective (Apu & Gavrilova, 2006). Recently, a growing number of studies were devoted to the swarm phenomenon inquiring the many possibilities of computational collaborations. For instance, self organizing swarm robots can potentially accomplish complex tasks and thus expensive and complex robotic intelligence may be replaced by rather unsophisticated devices. The principles of swarm have not only contributed to the field of Artificial Intelligence (AI), but also have been applied to the field of Virtual Reality (Raupp & Thalmann, 2001).

The concept of Swarm Intelligence is of particular interest in robotics because it would enable a manufacturer to produce cheap and expandable bots which can achieve many complex industrial tasks. Although not as common, the study of SI in tactical scenarios has been also explored in the literature. As an example, one of the projects conducted at BT Laboratory, University of Calgary, presented a unique approach that combines Swarm Intelligence with Genetic Algorithms by employing a bisexual mating process implemented using a Genetic Algorithm (GA) similar to the MOEO (Multi-Objective Evolving Object) method (Apu & Gavrilova, 2012). It also examines the relationship between formation and the tactical fitness of social agents. Therefore, in addition to answering how various formations are formed using limited intelligence, the research investigated why such formations are necessary for a swarm's global survival and resulting tactics developed.

Another aspect of intelligent computing is a neural network concept. The idea originated from observations on how human brain establishes connection among neurons as learning process takes place. In the presence of high-dimensional complex information, learning most significant patterns, associations and relationships between entities in the real word or computer world is of essence. *Chaotic Neural Networks* (CNNs) allow the learning process to take place in efficient, accurate and verifiable manner, while allowing for changes in learning environment and for almost unlimited complexity of the learned domain. The methodology made its way into the biometric domain, where processes mimicking the human brain are applied for complex decision making in the presence of uncertainty and noisy data (Gavrilova & Ahmadian, 2011).

3. COMPUTATIONAL INTELLIGENCE AND BIOMETRICS

All of these approaches are examples of *computational intelligence*—a field of study that focuses on computer processes imitating or simulating human intelligence, in application to a variety of domains. This book examines the application of computational intelligence to one of most dynamic area of computer based security—*Biometrics*.

A *biometric identification system* is an automatic pattern recognition system that recognizes a person by determining the authenticity of a specific physiological and/or behavioral characteristic (biometric) possessed by that person (Yanushkevich, Gavrilova, Wang, & Srinhari, 2007). *Physiological biometric identifiers* typically include fingerprints, hand geometry, ear patterns, eye patterns (iris and retina), facial features and other physical characteristics. *Behavioral identifiers* are voice, signature, typing patterns and others. Controlling access to prohibited areas and protecting important national or public interests is one of the main mandates of security and intelligence services. Frequently, to decide if a person is allowed to access a prohibited area, a biometric system is used. But a system based solely on a single source of information may be hindered by higher error rates in comparison to the system which relies on more than a single source of information to make its decision. Moreover, no existing biometric system even with the most powerful algorithms can provide one hundred percept reliable results, especially in the presence of noisy, erroneous, or forged data.

Thus, *multimodal biometric system* has emerged as a powerful methodology that helps to alleviate the deficiencies of any single biometric. Multi-modal biometric system can incorporate more than one biometric characteristic, instance, algorithm or raw data sample to improve the overall performance. The performance is measured in recognition accuracy, memory requirements, security (resistance to attacks, spoofing, or forgery) and circumvention (ability to produce consistent results even in the presence of noisy or low quality data, or in the absence of some type of data all together) (Ross, Nandakumar, & Jain, 2006). Further gradation of parameters to study system accuracy must take place at recognition level analysis. Criteria such as false acceptance rate, false rejection rate, and their combinations are frequently sued to further evaluate biometric system potential in real-life deployment for security applications.

The advantages of multimodal systems stem from the fact that there are multiple sources of information. The most prominent implications of this are increased accuracy, fewer enrolment problems and enhanced security. All multimodal biometric systems need a *fusion module* that takes individual data and combines it in order to obtain the authentication result: impostor or genuine user. The decision making process in a fusion module may be as simple as performing a logical operation on single bits or as complex as intelligent system developed using principles of fuzzy logic and cognitive informatics (Wang, et al., 2011, Saeed & Nagashima, 2012).

Throughout the last decade, a large number of multimodal systems have been developed in a quest to find an optimum combination of biometric characteristics and fusion approaches to minimize the recognition errors. The drive to develop new methodologies based on a fusion of features, algorithms and decision-making strategies can on its own be considered an intelligent approach to biometric.

4. SUMMARY

To summarize, this book provides state-of-the-art approached and novel treatment to biometric security decision-making domain. It presents methods and approaches from the fields of pattern recognition, security, and image processing, enriched with concepts from the fields of information fusion, computational intelligence, robotic biometric, and neural networks. The concept of biometric and multimodal biometric is introduced in-depth in the first chapters. Specific emphasis on multi-modal rank information fusion and consumer applications in security is given next, where information fusion is discussed in detail. Rank level fusion and its

variant are given specific treatment due to their advantages for security domain. In this section, an in-depth treatment of multi-modal biometric system architecture is presented, and pros and cons of various information fusion methods are discussed. In the next chapter, novel approaches based on fuzzy logic concept and Markov chain method are introduced as alternative rank fusion mechanisms. Their performance is illustrated with numerous examples and experimentations. Finally, novel alternative approaches based on computational intelligence paradigm are applied for the first time to the field of multi-modal biometric. These include chaotic neural network and dimension-reduction concepts for multi-biometric system design, robotic biometric and avatar recognition for intelligent software security systems, and a concept of application of soft biometric, social networks and social trends for improved performance of multi-modal biometric system. Finally, the last chapter outlines promising research directions and takes a glimpse into what future might hold for this dynamic research domain.

REFERENCES

Ahn, L. V., Blum, M., Hopper, N., & Langford, J. (2003). CAPTCHA: Using hard AI problems for security. *Lecture Notes in Computer Science, 2656*, 294–311. doi:10.1007/3-540-39200-9_18.

Apu, R., & Gavrilova, M. (2006). Battle swarm: An evolutionary approach to complex swarm intelligence. In *Proceedings of the 9th International Conference on Computer Graphics and Artificial Intelligence*, (pp. 139-150). Limoges, France: Eurographics.

Apu, R. A., & Gavrilova, M. L. (2012). Battle swarm: The genetic evolution of tactical strategies and battle efficient formations. *ACM Transactions on Autonomous and Adaptive Systems*. Retrieved from http://3ia.teiath.gr/3ia_previous_conferences_cds/2006/Papers/Full/Apu24.pdf

Bonabeau, E., Dorigo, M., & Theraulaz, G. (1999). *Swarm intelligence: From natural to artificial systems*. Oxford, UK: Oxford University Press.

Craig, N. L., Cohen-Fix, O., Green, R., Greider, C. W., Storz, G., & Wolberger, C. (2011). *Molecular biology principles of genome function*. Oxford, UK: Oxford University Press.

Gavrilova, L., & Yampolskiy, R. V. (2011). Applying biometric principles to avatar recognition. *Transactions on Computational Science*, *12*, 140–158. doi:10.1007/978-3-642-22336-5_8.

Gavrilova, M., & Ahmadian, K. (2011). Dealing with biometric multi-dimensionality through novel chaotic neural network methodology. *International Journal of Information Technology and Management*, *11*(1/2), 18–34. doi:10.1504/IJITM.2012.044061.

Naor, M. (1996). *Verification of a human in the loop or identification via the turing test*. Rehovot, Israel: Weizmann Institute of Science.

Raupp, S., & Thalmann, D. (2001). Hierarchical model for real time simulation of virtual human crowds. *IEEE Transactions on Visualization and Computer Graphics*, *7*(2), 152–164. doi:10.1109/2945.928167.

Ross, A. A., Nandakumar, K., & Jain, A. K. (2006). *Handbook of multibiometric*. Berlin, Germany: Springer.

Saeed, K., & Nagashima, T. (2012). Biometrics and Kansei Engineering. Berlin, Germany: Springer.

Turing, A. M. (1950). Computing machinery and intelligence. *Mind*, *59*, 433–460. doi:10.1093/mind/LIX.236.433.

von Neumann, J. (1966). *Theory of self-reproducing automate*. Urbana, IL: University of Illinois Press.

Wang, Y., Berwick, R. C., Haykin, S., Pedrycz, W., Baciu, G., & Bhavsar, V. C. … Zhang, D. (2011). Cognitive informatics in year 10 and beyond. In *Proceedings of the 10th IEEE International Conference on Cognitive Informatics & Cognitive Computing (ICCI*CC)*. IEEE.

Yanushkevich, S., Gavrilova, M., Wang, P., & Srihari, S. (2007). *Image pattern recognition: Synthesis and analysis in biometrics*. New York, NY: World Scientific Publishers.

Zykov, V., Mytilinaios, E., Adams, B., & Lipson, H. (2005). Robotics: Self-reproducing machines. *Nature*, *435*, 163–164. doi:10.1038/435163a PMID:15889080.

Chapter 2
Overview of Biometrics and Biometrics Systems

ABSTRACT

Recent security threats increase the necessity to establish the identity of every person. Biometric authentication is a solution to person authentication by analyzing physiological or behavioral characteristics. In this chapter, various biometric notions and terms are reviewed, along with typical biometric system components and different functionalities and performance parameters. The design and development of a biometric system, depending on a particular application scenario, is covered. This chapter also focuses on the inherent issues associated with biometric data and system performance through introducing radically new methods based on intelligent information fusion and intelligent pattern recognition, thus creating a notion of intelligent security systems. At the end of the chapter, the potential drawbacks of biometric unimodal systems, which serves as the motivation to introduce the concept of multimodal biometric system in the context of intelligent security systems, is discussed.

1. INTRODUCTION

Controlling access to prohibited areas and protecting important government and civilian objects are among the main activities of national and international security organizations. Similarly, with the advancement of large-scale networks (e.g., social networks, e-commerce, e-learning) and the growing concern for identity theft problems, the design of appropriate personal authentication systems is becoming more and more important. Usually, person authentication for access control to a prohibited area or for identification in different networks or social services scenarios (e.g.,

DOI: 10.4018/978-1-4666-3646-0.ch002

banking, welfare disbursement, immigration policies, etc.) is done using biometric authentication. According to Ratha et al. (Ratha, Senior, & Bolle, 2001), "Biometrics is the science of identifying or verifying the identity of a person based on physiological or behavioral characteristics." Over the last decades, people are using biometric authentication system in lieu of password or token based authentication systems for properties such as uniqueness, permanence over time, universality, user acceptance, and ease of use (Jain, Boelle, & Pankanti, 1999).

2. BIOMETRIC IDENTIFIERS

Biometric authentication offers a natural and reliable solution to the problem of establishing identity of a person utilizing his/her physiological or behavioural biometric characteristics or identifiers (Jain, Flynn, & Ross, 2007). The term "biometry" literally means "life science," and focused on studying biometric identifiers. These biometric identifiers, also called biometric traits, are integral part of a person's identity (Bolle, Connell, Pankanti, Ratha, & Senior, 2004). Some of the *physiological* characteristics that are now used for biometric recognition include face, fingerprint, hand-geometry, ear, iris, retina, DNA, palmprint, hand vein, etc. Voice, gait, signature, keystroke dynamics are examples of *behavioral* characteristics used for biometric recognition. *Soft biometrics* emerged as a new group of biometric gaining more and more attention. It includes measurements related to person's height, race, age, and gender. Finally, we identify one more group: *social* biometrics, making its way into the state-of-the-art security systems. This group includes data obtained from observing social behavior of the subject, interests, social network connections, work and leisure patterns, hobbies, and communication over social media.

2.1. Physiological Identifiers

Physiological biometrics are based on the body measurements, where a basic method to obtain the data is direct measurement of a part of the human body (Biometrics, 2009). Generally, it is assumed that physiological biometric identifiers are more stables than behavioral identifiers because most physiological identifiers remain unchanged over the course of individual's lifetime and do not depend significantly on external factors (Kung, Mak, & Lin, 2005). Face, fingerprint, and iris are the most commonly used physiological identifiers in today's automatic authentication systems. The other physiological biometric identifiers include retina, DNA, hand-geometry, ear shape, palmprint, hand, vein, teeth. Figure 1 shows some of the physiological biometric identifiers used for person authentication.

Face: Face is the most widely used biometric identifier to conduct human authentication. It is used every day by nearly everyone as the primary means for recognizing other humans. Among all the biometric traits, face is the most common and heavily used biometric for person identification. Face recognition is friendly and non-invasive (Feng, Dong, Hu, & Zhang, 2004).

The advantages of facial recognition include high public acceptance of the modality, commonly available sensors, not physically intrusive nature, and the ease with which humans can verify the results of security system based on facial biometric (Wilson, 2010).

Challenges in the face recognition process include different illumination conditions and backgrounds, changes in the facial expressions, aging, camouflage, and occlusions of some facial features (Singh, 2008). These challenges may reduce the overall recognition accuracy if not properly addressed.

Fingerprint: Fingerprint is one of the first biometric identifiers being used for recognition of one's belongings. Merchants in ancient times

Figure 1. Physiological biometric identifiers

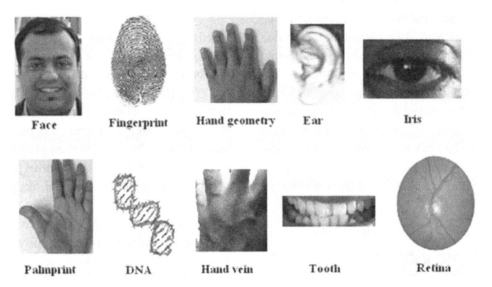

have been known to mark their goods (such as clay pots) with their palm or finger prints to identify the ownership. Most prevalent fingerprint characteristic is its uniqueness and consistency over time.

A fingerprint structure consists of a number of ridges (upper skin layer segments) and valleys (lower skin layer segments) on the surface of the finger: different minutia points such as ridge endings (where a ridge end) and ridge bifurcations (where a ridge splits into two) are made of by ridges (Biometric News Portal, 2012). Other characteristics such as topology of ridges and distances between them can be also considered as differentiating features.

Fingerprint structure is unique, and dictated by the ridges and the minutiae points relative topological positions (Jain, Flynn, & Ross, 2007). The main technologies for fingerprint sensing are ink-based, optical, and ultrasound. Two algorithms—minutia matching and topology matching, are generally used to recognize the fingerprints. Minutiae matching compares specific details within the fingerprint ridges and topology matching compares the overall topological structure of the fingerprints, and can be based on independent

distance or vector comparisons, or geometric structures similarities, such as triangulations of Voronoi diagrams (Wang & Gavrilova, 2006).

Iris: The iris is a ring that surrounds the pupil of an eye and has a muscular structure reacting to light entering an eye (Vacca, 2007). An iris recognition system processes the input image to extract the biometric features of the iris. These features are then stores and later compared with the purpose of iris identification or verification (Iris Recognition, 2003).

Iris pattern recognition is generally considered to be the most accurate among all the biometric traits available today, with the exception of, perhaps, only the DNA. However, the difficulty in obtaining DNA samples and the high cost of their processing make iris one of the preferred choices of companies looking for high level of security. As reported in Iris Recognition (2003), a combination of "high confidence authentication with factors like outlier group size, speed, usage/human factors, platform versatility and flexibility for use in identification or verification modes", makes iris recognition a highly versatile biometric which also can be used in a large-scale civil applications. Relative complexity in obtaining large iris database for method validation and security system's testing has led some researchers to focus on iris *synthesis*, i.e the inverse problem of biometric. The method is based on reverse subdivision and combining features from different irises to obtain a new instance (Wecker, Samavati, & Gavrilova, 2005). Iris is not the only biometric which can be reconstructed from studying and simulating common patterns—recent book on the subject: *Image Pattern Recognition: Synthesis and Analysis in Biometrics,* produced in 2007 by World Scientific Publishers, deals extensively with the subject and looks at data synthesis in variety of biometric areas (Yanushkevich, Gavrilova, Wang, & Srihari, 2007).

Hand Geometry: Hand geometry based biometric authentication systems use measurement based on the hand's shape and size, as well as the lengths and widths of the fingers. The main benefits of this kind of authentication methods rely on their simplicity and inexpensive implementation. However, hand geometry is much more similar and less discriminating between different individuals, as well as changes in size over one's lifetime (Biometric News Portal, 2012). Also, the size of sensing device is significantly larger than that of modern fingerprint scanners or cameras. Due to these reasons, these kinds of systems cannot be scaled for larger databases. However, these systems are being researched and employed on smaller scale for fraction of population.

Palm Print: Palm print-based biometric authentication systems compare individual palmprint to distinguish an individual. Palmprint contains less of the unique information available on the fingerprint, but has more details in terms of lines, wrinkles and creases (Swathi, 2011). It is often combined with hand shape biometric for better recognition accuracy. Palm print based personal verification has being an active research domain over the past 20 years, with interest slowly shifting towards other biometrics such as ear shape or retina.

Ear Shape: Ear is not a frequently used biometric trait, but is useful since the ear anatomy is unique to each individual and that features based upon measurements of that anatomy are unchangeable over time. Given that ears are unique, ear biometrics is worth investigations due to non-invasive and relatively unique characteristics (Burge & Burger, 1998).

Ear images can be acquired in a similar manner to face images (i.e., the camera which is used for acquiring face can also be used for acquiring ear images) and can be efficiently used in surveillance scenarios. The unique characteristic of ear biometric is that both image based and signal based processing methods can be used for their comparison.

Retina: Retina is the pattern of veins beneath the back of the eyeball, and is believed to be unique to each person (Jain, Flynn, & Ross, 2007). As a common sensing method, a low intensity beam or infrared light is projected into the eye so that a part of the retinal structure can be digitized (Griaule Biometrics, 2012). Unlike facial or ear shape biometric, since the image acquisition process requires cooperation of the subject, it is not a well researched nor well accepted method.

DNA: Deoxyribonucleic Acid (DNA) is a core biological code of all living organisms (Griaule Biometrics, 2012). DNA could be considered as one of biometrics since it can identify individuals uniquely. It has been the method of choice in a number of law enforcement investigations and has been accepted as evidence in courts. Except for the identical twins (who share the same DNA patterns), every individual has unique DNA patterns. The basis of DNA identification is the comparison of alternate forms of DNA sequences found at identifiable points in nuclear genetic material (Bolle, Connell, Pankanti, Ratha, & Senior, 2004).

Three factors have been mentioned in Griaule Biometrics (2012) which are the basis of the low degree of popularity of this biometric characteristic: (1) privacy concerns, since some additional information of the individual could be inferred from this data (such as genetic code, diseases, etc.), (2) real-time authentication capabilities, as this technique involves high computational resources and is difficult to be automated due to required chemical process, and (3) access availability, since it is easy to steal a piece of DNA from an individual and this information could be therefore used for fraudulent purposes (Griaule Biometrics, 2012).

Hand Vein: Vein pattern is the network of blood vessels beneath person's skin. The idea of using vein patterns as a form of biometric technology was first proposed in 1990s, with researchers paying increasing attentions to vein authentication over last twenty years (Tanaka & Kubo, 2004). Similarly to hand geometry, vein patterns are not unique, but are resistant to changed due to aging or trauma.

While there is inconclusive evidence, it seems that the patterns of veins are unique to each individual, including twins (Biometric News Portal, 2012; Griaule Biometrics, 2012). Interestingly enough, the back of a hand or a palm has a complex vascular pattern and thus can be compared to a fingerprint in its differentiating ability and a large number of features (Shrotri, Rethrekar, Patil, Bhattacharyya, & Kim, 2009).

Tooth: Tooth biometric is also called a dental biometrics, and it uses dental radiographs for human identification. The dental radiographs contain information about tooth contours, positions of neighboring teeth, and types of the dental procedures done on an individual (Chen & Jain, 2005).

This is not a popular biometric identifier due to large intra class variations and the difficulty in acquiring dental information. It is also is not as acceptable as face biometric or even a fingerprint. Hence, tooth-based authentication can be used as a supplementary biometric technique.

2.2. Behavioral Identifiers

Behavioral biometric identifiers are associated with persons' behavioral patterns. There are unique ways of how people do things such as walking, talking, signing their name, or typing on a keyboard (speed, rhythm, pressure on the keys etc.). The biometric identifiers that are associated with these behavior patterns are gait, voice, signature, and keyboard typing patterns, among which voice and signature are the most commonly used behavioral identifiers used in biometric authentication systems. These biometrics traits are non-contact but have high variations which are difficult to handle in some application environments.

While behavioral biometrics can be difficult to measure due to various influences such as stress, fatigue, or illness, they are sometimes more acceptable to users and generally cost less to implement (Kung, Mak, & Lin, 2005). For example, one of the widely used behavioral biometric systems is the speaker recognition system which uses individual voice for authentication. Figure 2 shows some behavioral biometric identifiers that are used in biometric based person authentication system.

Voice: Although voice is a combination of physical and behavioral characteristics that are related to the voice signal patterns of a given individual, it is usually considered as behavioral identifier. Voice is one of the most widely used biometric identifiers along with face and fingerprint to recognize individuals by how they sound when speaking.

The physical characteristics of voice are related to the way the humans form the sound. These characteristics include the vocal tracts, mouth, nasal cavities, and lips (Griaule Biometrics, 2012). The behavioral characteristics of voice can be affected by the emotional states of the speaker (Griaule Biometrics, 2012). Voice-based authentication methods have been mainly developed in the context of voice-recognition commercial systems and language translators, with intermediate success up to date.

Signature: The handwriting of a given individual can be thought as representing his/her own characteristics. Signatures have been used before the advent of computers in different areas

Figure 2. Behavioral biometric identifiers (image sources: Google)

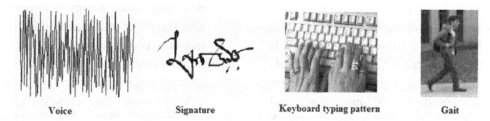

| Voice | Signature | Keyboard typing pattern | Gait |

ranging from government and legal departments to commercial and business applications (Griaule Biometrics, 2012).

Traditionally, signature authentication in a security system may be either static or dynamic. Static signature authentication uses only the statistical features of the signatures, whereas the dynamic authentication uses not only those features, but also some additional information such as velocity, acceleration, pressure, and trajectory of the signatures. Signatures have large intra class variations due to the physical and emotional state of a person, and may vary over a period of time. However, such systems may be incorporated transparently since individuals are used to provide their signatures in different environments of their daily life (Biometric News Portal, 2012).

Gait: Gait is a behavioral characteristic used to authenticate people by the way they walk (Griaule Biometrics, 2012). The strength of gait recognition is in its applicability to recognition of people at a distance in video images and thus suitable for surveillance systems. Further, there is no need of big co-operation efforts from the subject and hence can be effectively used in airport or other crowded places. Therefore, gait has received a lot of attention from studies in medicine, psychology, and human body modeling (Jain, Flynn, & Ross, 2007). With the vast availability of video surveillance data, there is also a variety of methods for extracting gait patterns from video sequences and their subsequent analysis.

Keyboard Typing Pattern: Keyboard typing pattern or, in other words, keystroke dynamic is related to the way people type characters on keyboards (360 Biometrics, 2012). Each person has a unique timing between key strokes and the hold time between keys, which is the basis for typing pattern based behavioral biometric across the population in general (Bolle, Connell, Pankanti, Ratha, & Senior, 2004). One of the main benefits of this technique is that it allows 'continuous authentication', since the individual can be analyzed over a large periods of time.

2.3. Soft Biometric Identifiers

The new group of biometric gaining more attention is *soft* biometrics. It includes measurements related to person's observable information such as height, race, age, gender, hair, or eye color. This information is typically stored in government issued identification documents (i.e. passport, driver's license), and can be covertly obtained (i.e. without person physical contact or notice) and verified against person's record stored in centralized database. Utilizing this information can offer clues in case when main biometric system modules are not operational, or in low security application scenarios.

2.4. Social Biometric Identifiers

Social biometrics is an emergence area of research pursues at BTLab at the University of Calgary. It has been recently making its way into the state-of-the-art security systems. This type of biometric includes data obtained from observing social behavior of the subject, interests, social network connections, work and leisure patterns, hobbies, and communication over social media. While more complex to obtain and analyze, this type of biometric data includes invaluable information for profiling a subject. As in a case of twins, even as physical appearance can be identical, the social connections, interests, and contacts would vary significantly. The choice of what biometric to use in a security system depends on a number of factors, including purpose of the system, its scope, number of users, budget, social environment, usability studies, personnel training, mode of operation, duration of active deployment and particular application scenario.

3. ATTRIBUTES OF BIOMETRIC IDENTIFIERS

Each of the biometric identifiers discussed above has its own advantages and disadvantages. Thus, based on the application scenarios, biometric system uses one or more biometric identifiers taking into consideration a number of factors. Researchers have identified several requirements that a biometric identifier needs to possess in order to be used in an authentication system (Jain, Boelle, & Pankanti, 1999). These requirements are either theoretical or practical. Theoretical requirements include (Jain, Boelle, & Pankanti, 1999):

- **Universality:** Each person in the population should have the biometric identifier.

- **Distinctiveness:** Identifiers for two persons randomly selected from the whole population should be sufficiently different across individuals comprising the population.
- **Permanence:** The biometric identifiers should remain the same over a period of time or should change relatively slowly.
- **Collectability:** The identifier should be able to be acquired, digitized and stored using appropriate devices.

The practical requirements are related to the functionality of the biometric systems (Jain, Boelle, & Pankanti, 1999):

- **Performance:** The achievable recognition accuracy, speed or other important parameter.
- **Acceptability:** The acceptance of the end-users of the biometric system in their daily lives.
- **Circumvention:** The degree of security of the system to noise or attacks.

4. COMPONENTS OF BIOMETRIC SYSTEM

The word 'Biometric' is a composite of two parts: the Greek words 'bios' (life) and 'metron' (measure) (Werner, 2008). Biometric is sometimes defined as a research area focused on measuring and analyzing a person's unique characteristics (Maltoni, Maio, Jain, & Prabhakar, 2009). Biometric systems are becoming increasingly popular due to some factors including increased need for reliable and convenient authentication, decreased costs and increased government and industry adoption. With biometric based authentication, there is nothing to lose or forget unlike with physical tokens (keys, cards) or information tokens (PIN, code) in traditional security authentication systems. In addition, the cost of biometric systems was brought down to an affordable range at the commercial market exhibits continuous improvement in the hardware and software technologies and accessibility. Due to these advantages, numerous public and private organizations are using biometric systems as main security systems for access control based on person authentication. Table 1 (Jain, Boelle, & Pankanti, 1999) shows comparison among different types of biometric identifiers:

A choice of particular identifier is thus belongs largely with system architect and depends on variety of requirements on performance, cost, accessibility, training, deployment, and system maintenance. A typical biometric system operates by acquiring biometric data from an individual, extracting a feature set from the acquired data, and comparing this feature set against the template feature set in the database. Thus, biometric system components can be divided into some modules according to their functionalities. These modules are typically sensor or data acquisition module, feature extraction module, matching module and decision module (Jain, Flynn, & Ross, 2007).

4.1. Sensor or Data Acquisition Module

This is the first step of any biometric system where biometric data is acquired from the source, i.e., from an individual, through a variety of instruments or sensors such as camera, fingerprint sensor, microphone etc. The user's characteristics must be presented to a sensor in either cooperative (with user's agreement) or non-cooperative (remotely observable) fashion. The output of the data acquisition process becomes an input data (in a form of an image or a signal). Biometric data acquisition can be affected by human factors, such as training, experience or fatigue, environmental conditions expressed in the form of weather, light, sound interference and so on, quality and type of sensor used and cooperation of the end user. The Failure to Enroll Rate (FTE) typically measures the lack of success of this data acquisition process. Figure 3 shows some sample input images and signal obtained from face, signature, fingerprint, iris, and voice recognition systems.

4.2. Feature Extraction Module

The feature extraction module uses image processing or signal processing methods to obtain *biometric features*—a subset of high-dimensional data source which can be compared (matched) in a faster way than to compare the whole image. According to Ross, Nandakumar, and Jain (2006), features should be "unique for each person (i.e. display extremely small inter-user similarity) and invariant with respect to changes in the different samples of the same biometric trait collected from the same person (extremely small intra-user variability)." The initial task of feature extraction module is to perform pre-processing of the acquired data. Such

Table 1. Comparison among different types of biometric identifiers (Jain, Boelle, & Pankanti, 1999)

Biometrics	Universal	Unique	Permanence	Collectable	Performance	Acceptability	Circumvention
Face	High	Low	Medium	High	Low	Low	High
Fingerprint	Medium	High	High	Medium	High	Medium	High
Iris	High	High	High	Medium	High	Low	High
Signature	Low	Low	Low	High	Low	High	Low
Voice	Medium	Low	Low	Medium	Low	High	Low
Vein	Medium	Medium	Medium	Medium	Medium	Medium	High
DNA	High	High	High	Low	High	Low	Low

Figure 3. Sample input and acquired signals for a biometric system

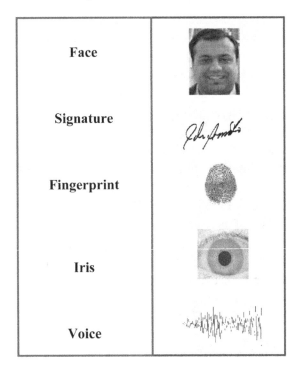

pre-processing might include image binarization, normalization, or segmentation. The goal of this process is to simplify the raw data and change the representation of an image or signal into something that is more meaningful and easier to analyze. For example, in an iris recognition based biometric authentication system, segmentation is required to isolate the iris region from input eye images. The segmented portion of the input data is then further processed to extract meaningful features. The feature set obtained by this module is then stored in the system database as a template.

4.3. Matching Module

The matching module is one of the key modules that compares the feature set extracted from biometric sample during feature extraction to the template stored in the biometric database. The matching module determines the degree of similarity or dissimilarity between the sample

and the template. Such a procedure can rely on a variety of measures, based on Euclidean, Minkowski, Lyapunov, Mahalanobis, Geodesic, or other distance-based methods. It can also rely on comparing distance between vectors in PCA (Principal Component Analysis) eigenface-based face recognition, or between clusters measure in Chaotic Neural network method. These similarity or dissimilarity scores are then given to the final module for the authentication decision.

The significant amount of literature devoted specifically to development of better algorithms used in this module. Their variability among different biometric systems is very high. These differences depend on the design of the application and are based on a variety of factors, including the application scenarios, time requirements and the availability of resources. Some heavily used algorithms used in this module are neural network, principal component analysis, support vector machines and fuzzy logic (Paul, Monwar, Gavrilova, & Wang, 2010; Gavrilova & Ahmadian, 2011).

4.4. Decision Module

The final module of a biometric system is the decision module. The authentication decision is taken at this module based on this degree of similarity or dissimilarity between features. The application requirements are taken into consideration in this module to select the final decision. As an example, given an 88% match, it can be treated as a positive decision in one application scenario, whereas the same matching rate might be considered as a negative decision in a more robust application scenario.

Two other important components of all biometric systems are the biometric information database and the communication channel. The biometric information database (can also be referred as system database) contains and manages all the extracted feature sets (templates). This component is accessed by matching module for comparison of the input feature sets with the templates.

The communication or transmission channel refers to the communication path between other components or modules. This channel is internal to the device in some self-contained systems and distributed in other systems. It consists of one central data storage with many remote data acquisition points. If a large amount of data is involved, in the latter scenario, data compression may be required before sending the data to the transmission channel or to the storage to conserve bandwidth and storage space.

Further, in some biometric systems, a quality check module is incorporated after the sensor module into the system to ensure the quality of the acquired biometric samples (Jain, Flynn, & Ross, 2007). If the acquired biometric samples do not satisfy the required criteria, they are re-acquired from the subject.

Figure 4 shows standard components of a typical biometric system.

4.5. Intelligent Security Systems

Due to increasing demands on performance of security system under various conditions and operational scenarios, alternative approaches has been recently developed to improve feature extraction, augment learning process and make better aggregate decisions. The overall theme, visible in these state-of-the-art approaches, is direction towards more intelligent and adaptive systems. Here, intelligent means independent, self-learning system, which can observe patterns, adapt to changes in environment or incoming data samples, and compensate for lapses related to imperfect technology or operator training, variability of user samples or inconsistency in data quality. As such, fuzzy logic methodology can be used instead/as part of decision module, or dimensionality-reduction module can be employed to reduce number of features extracted by the system. Intelligent learner module based on evolutionary computing, intelligent pattern recognition engine or neural network can be used to augment system recognition rate and for extensive training on given databases before active system deployment (Paul, Monwar, Gavrilova, & Wang, 2010; Gavrilova & Ahmadian, 2011). These methods will be reviewed in subsequent chapters of this book.

5. BIOMETRIC VERIFICATION

A biometric system can be used for person *verification* or person *identification*. Person verification answers the question, "Am I who I claim to be?" and then confirms the validity of a claimed identity by comparing a verification template to an enrollment template (Jain, Flynn, & Ross, 2007). Verification thus needs a person to provide his identity in order for it to be verified. Thus, the comparison needed for verification is called *one-to-one* comparison (Jain, Flynn, & Ross, 2007). During verification, usually some knowledge about the identity (such as ID) is given to the system along with the biometric identifier. This additional factor uniquely presents an enrolled identity and extracted biometric features to the system database (Bolle, Connell, Pankanti, Ratha, & Senior, 2004). Verification is used in everyday life at such circumstances as banking, using credit card, attending events, taking exams and so on. Usually, a person's identity is verified by the means of comparing his facial biometrics and/or signature to the data stored on his passport, ID or credit card. Sometimes, more than one source of information is used for such verification.

6. BIOMETRIC IDENTIFICATION

Biometric identification establishes a person's identity by answering the question "Who am I?" To do so, an identification system performs matches to test person's identity against multiple biometric templates. Thus, in identifications

Figure 4. Block diagram of a typical biometric system

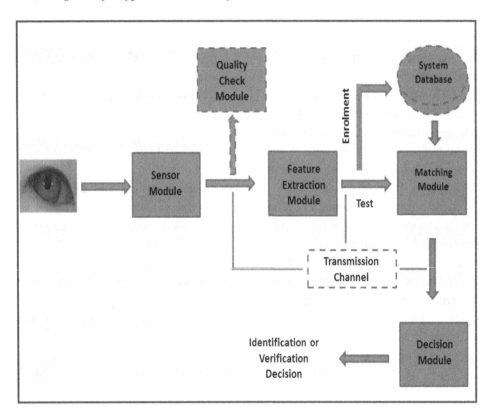

system, matching is *one-to-many* matching (Jain, Flynn, & Ross, 2007).

There are two types of identification systems: positive identification and negative identification (Jain, Flynn, & Ross, 2007). The goal of a positive identification system is to locate a user's biometric information in a biometric database. A common example used in the literature of a positive identification system is an inmate identification, where a camera for capturing a face or an iris is used instead of a usual ID.

Negative identification systems compare one template against many, but their goal is to verify that a person is not enrolled in the database (Jain, Flynn, & Ross, 2007). This can prevents people from enrolling twice in a system, and is often used in social services programs in which users may be tempted to enroll multiple times to gain extra benefits (Jain, 2005).

There is also a separate category between identification and verification, called one-to-few matching (Bolle, Connell, Pankanti, Ratha, & Senior, 2004). Here, few means that the database size is really small, however in practice such classification is rarely found to be useful.

For both verification and identification, successful biometric enrollment is necessary. Biometric enrollment is the process of registering subjects in biometric databases (Jain, Flynn, & Ross, 2007). In this process, the user's biometric data is captured, preprocessed and features extracted as shown in Figure 4. The user's template is then store in the system database. Figure 5 illustrates biometric enrollment, biometric verification, and biometric identification processes.

Figure 5. Biometric enrolment, biometric verification, and biometric identification

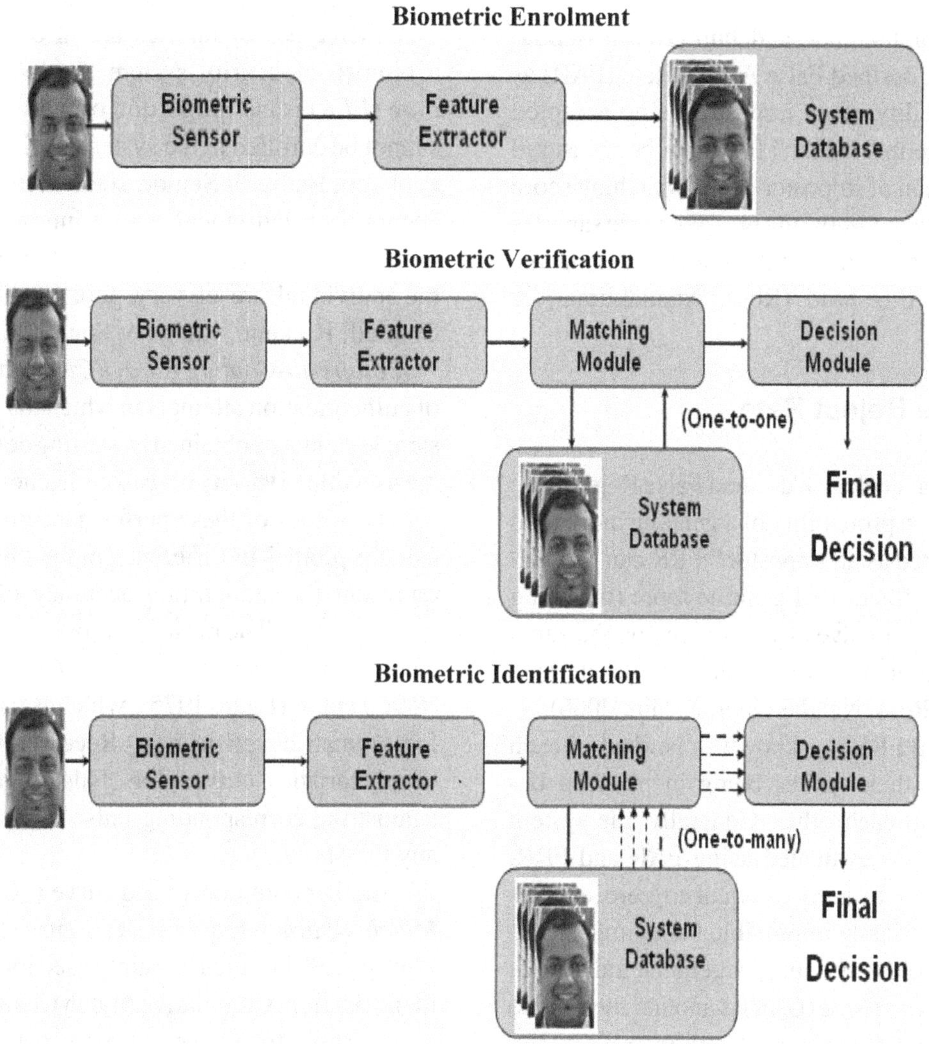

7. BIOMETRIC SYSTEM PERFORMANCE

Performance of a biometric system is usually expressed by some parameters. A decision made by a biometric system is either a "genuine individual" type of decision or an "impostor" type of decision (Ross, Nandakumar, & Jain, 2006). For each type of decision, there are two possible outcomes, true or false. Therefore, there are a total of four possible outcomes: a genuine individual is accepted or a genuine match occurred, a genuine individual is rejected or a false rejection occurred, an impostor is rejected or a genuine rejection occurred and an impostor is accepted or a false match occurred (Ross, Nandakumar, & Jain, 2006). As with any process, the correctness of the procedure can be estimated in a formalized manner through special parameters, namely two error rates—false accept rate and false reject rate, often referred to as false acceptance rate or false rejection rate (with the same abbreviation sued in both cases).

7.1. False Accept Rate

In Ross, Nandakumar, and Jain (2006), Nandakumar et al. defined False Accept Rate (FAR) as "the probability of an impostor being accepted as a genuine individual." FAR can be measured as the fraction of impostor score (matching score which involves comparing two biometric samples originating from different users) exceeding the predefined threshold (Ross, Nandakumar, & Jain, 2006).

7.2. False Reject Rate

Nandakumar et al. also defined False Reject Rate (FRR) "as the probability of a genuine individual being rejected as an impostor." FRR can be measured as the fraction of genuine score (matching score, which involves two samples of the same biometric trait of a user) below the predefined threshold (Ross, Nandakumar, & Jain, 2006).

FAR and FRR are known to be dual of each other, with their values being in inversed dependence on each other. Generally, the system performance is evaluated using FAR and FRR. A FAR close to zero or equal to zero ensures that it is practically impossible for an imposter to get access to the system as a genuine individual. *Genuine Accept Rate (GAR)* is another measure to evaluate the performance of a biometric security system. It is defined as the "fraction of genuine score exceeding the predefined threshold" (Ross, Nandakumar, & Jain, 2006). As generally agreed in the biometric community, the following equation is used to compute the Genuine Accept Rate of a system:

$$GAR = 1 - FRR \qquad (2.1)$$

Genuine Reject Rate (GRR) is the "fraction of impostor score below the predefined threshold" (Ross, Nandakumar, & Jain, 2006). The following equation can be used to find out the Genuine Reject Rate of a system:

$$GRR = 1 - FAR \qquad (2.2)$$

Other types of failures are also possible in a biometric security system. *Failure-to-Enroll Rate (FER)* is the proportion of individuals who cannot be enrolled in the system (Bolle, Connell, Pankanti, Ratha, & Senior, 2004). This error can occurs if an individual cannot interact with the biometric system or if the biometric samples of the individual are of very poor quality (Bolle, Connell, Pankanti, Ratha, & Senior, 2004).

Failure-to-Capture Rate (FCR) is the fraction of authentication attempts in which the biometric sample cannot be obtained by sensing devices from the user due to variety of reasons (Renesse, 2002).

The values of these performance metrics are usually plotted in different graphs or curves to represent the recognition accuracy of the biometric system. The most commonly used plotting curve is the *Receiver Operating Characteristics (ROC)* curve (Egan, 1975), which is used mostly for biometric verification. Receiver Operating Characteristics curve plots False Accept Rate against the corresponding False Reject Rate for any threshold.

Another commonly used curve is *Cumulative Match Characteristics (CMC)* curve (Moon & Phillips, 2001) which is mainly used for biometric identification. Cumulative Match Characteristics curves show the chance of a correct identification within the top ranked match results. A good system will start with a high identification rate for low ranks identities (Dunstone & Yager, 2006).

The performance of a biometric system may also be expressed using *Equal Error Rate (EER)* and *d-prime* value. The Equal Error Rate refers to that operating point at the intersection of the line FAR = FRR with the ROC of the matcher. The Equal Error Rate is the value of the error rates at this point EER = FAR = FRR.

In Figure 6, the Equal Error Rate EER_x of matcher x is clearly less than the Equal Error rate EER_y of matcher y. d-prime is another way to judge the quality of the matcher (Bolle, Connell,

Figure 6. Equal error rate (EER) comparison of two separate sample systems (Bolle, Connell, Pankanti, Ratha, & Senior, 2004)

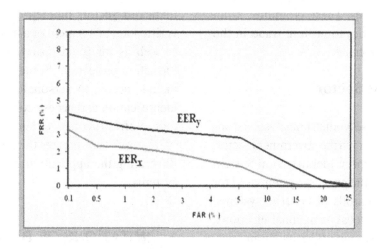

Pankanti, Ratha, & Senior, 2004). Daugman and Willaims (1996) suggested the following formula to find out the d-prime of a matcher which is denoted by.

$$d' = \frac{\mu_m - \mu_n}{\sqrt{\left(\sigma_m^2 + \sigma_n^2\right)}} \qquad (2.3)$$

8. APPLICATIONS OF BIOMETRIC SYSTEM

Due to the recent security threats originating from variety of sources (international terrorism, organized crime, commercial espionage, illegal immigration, cybersecurity, etc.), the use of biometric systems for person authentication has increased considerably as seen in variety of application scenarios including forensic, civil, and commercial sector, government sector, genetics, and health care.

8.1 Forensic

The use of biometric in law enforcement and forensic is known for at least a few hundred years. Fingerprint identification system is one of

the original and the most widely used biometric system for this purpose (Jain & Kumar, 2012). Such system is used not only to link suspects to crime scenes, but also to link persons arrested under another name to other potentially relevant cases, to identify victims of the crime and to associate persons with events in complex databases (Wayman, Jain, Maltoni, & Maio, 2005). Other biometric identifiers which are used in forensic sector include face, signature, gait, voice and DNA. Face and gait can be used for surveillance purposes very busy places such as stadiums, airports, meetings, etc. Signature and voice van be used for identification of criminals. Recently, DNA matching is used more and more to identify criminals due to a higher accessibility of such technology and its lowering costs.

8.2 Civil and Commercial Sector

The civil and commercial sector has been a major supporter, deplorer, and user of various biometric technologies for years (Woodward, Horn, Gatune, & Thomas, 2003). The application areas of biometric technology in this sector are social services, banking and financial services, time and attendance monitoring, E-commerce, E-learning and more recently virtual reality. Almost all of the

United State states and most of the Provinces in Canada use large-scale biometric applications in social service programs. These programs, as an example, use biometrics to prevent fraud in the form of multiple enrolments.

8.3 Government Sector

Government offices use biometric analysis and application of the biometry in the government sector. An Automatic Fingerprint Identification System deployed in States is the primary system used for locating duplicates enrolls in benefits systems, electronic voting for local or national elections, driver's license emission, etc. (Griaule Biometrics, 2012). The typical applications include national identification cards, voter ID and elections voter authentication, Driver's licenses, social benefits distributions, employee authentication and military programs. In almost all of these programs, the idea is to include digital biometric information in the identification cards. These programs must deal with large-scale databases, containing hundreds of millions of samples, corresponding to the large portions of population of one country. Traditionally, those applications are primarily based on finger-scan and fingerprint identification technology, with more and more systems that rely on a facial-scan and iris-scan technology. Finally, biometric passports often are based on a combination of ID, signature, fingerprint, and face, thus providing a higher level of security.

8.4 Genetics

Fingerprint pattern characteristics have been used for long time to trace the genetic history of population groups (Cummins & Kennedy, 1940). There is also research which associates certain fingerprint characteristics with certain birth defects and diseases (Woodward, 1997). There is, without a doubt, a vast number of applications of such research in medicine and public health.

8.5 Health Sector

In health sector, biometrics used for identity verification of a patient or a health-care provider, as well as for fraud prevention and patient information protection. Some typical applications include access to personal information, patient identification and access control to physical and digital infrastructures (Griaule Biometrics, 2012).

Figure 7 summarizes the discussion above by illustrating the application domain of biometric systems.

9. LIMITATIONS OF BIOMETRIC SYSTEM

In recent years, biometric systems have been successfully deployed in a number of real-world applications with some biometrics offering reasonably good overall performance. However, even the most advanced biometric systems to date are facing numerous problems, some inherent to the type of data and some of them inherent to system design. In particular, biometric systems generally suffer in person authentication process due to the factors listed below.

Noisy Data: Noise can be defined as unwanted data without meaning associated with the data. Noisy data is one of the common problems of biometric systems. Noise can be included in the biometric data during acquisition due to defective, improperly maintained or outdated sensors, due to the failure of providing noise free biometric data acquisition environment or simply produced as an unwanted by-product of other activities (Jain, 2005). For example, capturing voice biometric data in a noisy environment (i.e. during heavy rain, etc.) will result in a noisy voice signal enrolment. While an undesirable trend, the recognition accuracy of a biometric system could be sensitive to the quality of the biometric data (Garris, Watson, & Wilson, 2004). Developing better algorithms,

Figure 7. Application domains of biometric systems

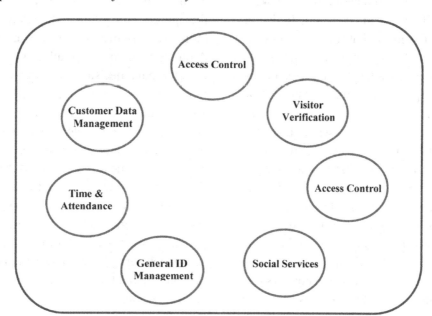

which can adapt to less than perfect input data and training the biometric system on varied in quality data may alleviate the problem.

Non-Universality: Universality is one of the most important requirements for a biometric trait. A biometric trait is said to be universal if all members of the target population can be enrolled in the biometric system (Chen, Dass, & Jain, 2005). It is easy to show that not all biometric users possess this property. For example, a blind person cannot present his/her iris or retina in front of the sensors or an illiterate individual cannot provide signature for biometric authentication.

Lack of Individuality: This problem occurs in biometric systems when the extracted feature sets of two different individuals are quite similar. For example, the facial appearance of a father and a son can be quite similar. This can limit the discrimination capability of a face-based biometric authentication system (Golfarelli, Maio, & Maltoni, 1997). False recognition rate of biometric can get higher than desired due to this lack of discrimination (Jain, 2005).

Intra-class Variation: This is the problem where two feature sets (for enrollment and for authentication) acquired from an individual are not identical. This can occur due to any sensor related issue or due to changes in the environmental conditions and inherent changes in the biometric trait. Large intra class variations usually decrease the recognition accuracy of the system (Uludag, Ross, & Jain, 2004).

Susceptibility to Circumvention: This problem occurs when an impostor presents a fake biometric sample to the system. Although it is very difficult to steal someone's biometric traits, studies (Matsumoto, Matsumoto, Yamada, & Hoshino, 2002; Putte & Keuning, 2000) have shown that it is possible to construct gummy fingers using lifted fingerprint impressions and utilize them to circumvent a biometric system. Behavioral traits like signature and voice are more susceptible to such attacks than physiological traits (Jain, 2005).

Privacy: Privacy is another problem with biometric systems since a biometric trait is a permanent link between a person and his identity. The acquired biometric trait can be used in such a way which will threaten a person's right to privacy (Bolle, Connell, Pankanti, Ratha, & Senior, 2004). To ensure that this does not happen, a number of legal, personal and technical steps can be taken. On a legal level, it should be unlawful to use person's biometric information for any purposes other than those it was collected for. On a technical level, it is essential that such sensitive information is stored in a secure manner and only authorized persons have an access to it. On a personal level, it is important for end users to understand that any time a data is entered in a computer based system; there is a slight chance of it falling into the wrong hands. Thus, one should exercise a reasonable degree of caution and judgment in deciding when and how to share the sensitive information. Overall, ensuring data security in a biometric system is very important.

Just recently, a *template protection* has emerged as one of the ways to protect system users from their information being exposed. The idea is based on storing only a portion of biometric data in a database, ideally in an encrypted form from which it is impossible to recover the original data. Development of such methods, while challenging, is one of the future research directions in biometric domain (Jain & Kumar, 2012).

10. SUMMARY

This chapter provides reader with an overview of various biometric notions and terms, specifically physical and behavioral identifiers. Among the biometric identifiers, face, fingerprint, signature, voice and iris are the most used biometric identifiers due to the ease of their availability and their recognition performance. New categories of biometric identifiers, namely soft biometrics and social biometrics, are also introduced.

This chapter also discusses biometric functionalities and performance parameters. Biometrics authentication can be subdivided into verification and identification. Due to the nature of the application, the user or developer must decide on the appropriate system architecture. The typical biometric system architecture is introduced next. Along with known modules, new approaches to intelligent decision-making, feature extraction, pattern recognition, and system learning are outlined. The overall trend over recent few years was to compensate for inherent issues with biometric data and system performance through introducing radically new methods based on intelligent information fusion and intelligent pattern recognition, thus creating a notion of intelligent security systems. At the end of the chapter, the potential drawbacks of biometric unimodal system which serves as the motivation to introduce the multimodal biometric system concept in the context of intelligent security systems has been discussed. Last but not the least, issues of privacy and security are given consideration, with a new approach to biometric template protection—cancellable biometric—identified as one of the highly active research directions in a biometric domain.

REFERENCES

Biometric News Portal. (2012). *Website.* Retrieved from http://www.biometricnewsportal.com/biometrics_benefits.asp

Biometrics. (2009). *Emerging devices technical brief.* New York, NY: AT&T.

Biometrics. (2012). *Website.* Retrieved online from http://360biometrics.com/faq/Keystroke_Keyboard_Dynamics.php

Bolle, R. M., Connell, J. H., Pankanti, S., Ratha, N. K., & Senior, A. W. (2004). *Guide to biometrics.* New York, NY: Springer-Verlag.

Burge, M., & Burger, W. (1998). Ear biometrics. In Jain, A. K., Bolle, R., & Pankanti, S. (Eds.), *Biometrics: Personal Identification in Networked Society* (pp. 273–286). Norwell, MA: Kluwer Academic Publishers.

Chen, H., & Jain, A. K. (2005). Dental biometrics: Alignment and matching of dental radiographs. *IEEE Transactions on Pattern Analysis and Machine Intelligence, 27*(8), 1319–1326. doi:10.1109/TPAMI.2005.157 PMID:16119269.

Chen, Y., Dass, S. C., & Jain, A. K. (2005). Fingerprint quality indices for predicting authentication performance. In *Proceedings of Fifth International Conference on Audio and Video-Based Biometric Person Authentication (AVBPA)*, (pp. 373-381). Rye Brook, NY: AVBPA.

Cummins, H., & Kennedy, R. (1940). Purkinji's observations (1823) on fingerprints and other skin features. *The American Journal of Police Science, 31*(3).

Daugman, J. G., & Williams, G. O. (1996). A proposed standard for biometric decidability. In Proceedings of cardTechSecureTech, (pp. 223-224). Atlanta, GA: cardTechSecureTech.

Dunstone, T., & Yager, N. (2006). *Biometric system and data analysis: Design, evaluation, and data mining.* New York, NY: Springer.

Egan, J. (1975). *Signal detection theory and ROC analysis.* New York, NY: Academic Press.

Feng, G., Dong, K., Hu, D., & Zhang, D. (2004). When faces are combined with palmprint: A novel biometric fusion strategy. In *Proceedings of First International Conference on Biometric Authentication*, (pp. 701-707). Hong Kong, China: IEEE.

Garris, M. D., Watson, C. I., & Wilson, C. L. (2004). *Matching performance for the US-visit IDENT system using flat fingerprints.* Technical Report 7110. Washington, DC: National Institute of Standards and Technology (NIST).

Gavrilova, M., & Ahmadian, K. (2011). Dealing with biometric multi-dimensionality through novel chaotic neural network methodology. *International Journal of Information Technology and Management, 11*(1-2), 18–34.

Golfarelli, M., Maio, D., & Maltoni, D. (1997). On the error-reject tradeoff in biometric verification systems. *IEEE Transactions on Pattern Analysis and Machine Intelligence, 19*(7), 786–796. doi:10.1109/34.598237.

Griaule Biometrics. (2012). *Website.* Retrieved online from http://www.griaulebiometrics.com/en-us/book/understanding-biometrics/introduction/types/behavioral

Iris Recognition. (2003). *Iris technology division.* Cranbury, NJ: LG Electronics USA.

Jain, A., & Kumar, A. (2012). Biometric recognition: An overview. *The International Library of Ethics. Law and Technology, 11*, 49–79.

Jain, A. K. (2005). Biometric recognition: How do I know who you are? *Lecture Notes in Computer Science, 3617*, 19–26. doi:10.1007/11553595_3.

Jain, A. K., Bolle, R., & Pankanti, S. (Eds.). (1999). *Biometrics: Personal identification in networked society.* Dordrecht, The Netherlands: Kluwer Academic Publishers.

Jain, A. K., Flynn, P., & Ross, A. (2007). *Handbook of biometrics.* New York, NY: Springer.

Kung, S. Y., Mak, M. W., & Lin, S. H. (2005). *Biometric authentication: A machine learning approach.* Upper Saddle River, NJ: Prentice Hall.

Maltoni, D., Maio, D., Jain, A. K., & Prabhakar, S. (2009). *Handbook of fingerprint recognition* (2nd ed.). New York, NY: Springer-Verlag. doi:10.1007/978-1-84882-254-2.

Matsumoto, T., Matsumoto, H., Yamada, K., & Hoshino, S. (2002). Impact of artificial 'gummy' fingers on fingerprint systems. []. SPIE.]. *Proceedings of SPIE Optical Security and Counterfeit Deterrence Techniques IV, 4677*, 275–289. doi:10.1117/12.462719.

Moon, H., & Phillips, P. J. (2001). Computational and performance aspects of PCA-based face recognition algorithms. *Perception, 30*(5), 303–321. doi:10.1068/p2896 PMID:11374202.

Paul, P. P., & Gavrilova, M. (2012). Multimodal cancellable biometric. In *Proceedings of the 10th IEEE International Conference on Cognitive Informatics & Cognitive Computing (ICCI*CC)*, (pp. 43-50). IEEE.

Paul, P. P., Monwar, M., Gavrilova, M., & Wang, P. (2010). Rotation invariant multi-view face detection using skin color regressive model and support vector regression. *International Journal of Pattern Recognition and Artificial Intelligence, 24*(8), 1261–1280. doi:10.1142/S0218001410008391.

Putte, T., & Keuning, J. (2000). Biometrical fingerprint recognition: Don't get your fingers burned. In *Proceedings of IFIP TC8/WG8.8 Fourth Working Conference on Smart Card Research and Advanced Applications*, (pp. 289-303). Bristol, UK: IFIP.

Ratha, N., Senior, A., & Bolle, R. (2001). Tutorial on automated biometrics. In *Proceedings of International Conference on Advances in Pattern Recognition*. Rio de Janeiro, Brazil: IEEE.

Renesse, R. L. V. (2002). Implications of applying biometrics to travel documents. []. Springer.]. *Proceedings of the Society for Photo-Instrumentation Engineers, 4677*, 290–298. doi:10.1117/12.462720.

Ross, A., Nandakumar, K., & Jain, A. K. (2006). *Handbook of multibiometrics*. New York, NY: Springer-Verlag.

Shrotri, A., Rethrekar, S. C., Patil, M. H., Bhattacharyya, D., & Kim, T.-H. (2009). Infrared imaging of hand vein patterns for biometric purposes. *Journal of Security Engineering, 2*, 57–66.

Singh, R. (2008). *Mitigating the effect of covariates in face recognition*. (PhD Dissertation). University of West Virginia. Morgantown, WV.

Swathi, N. (2011). New palmprint authentication system by using wavelet based method. *Signal & Image Processing: An International Journal, 2*(1), 191–203. doi:10.5121/sipij.2011.2114.

Tanaka, T., & Kubo, N. (2004). Biometric authentication by hand vein patterns. In *Proceedings of SICE Annual Conference*, (pp. 249-253). Sapporo, Japan: SICE.

Uludag, U., Ross, A., & Jain, A. K. (2004). Biometric template selection and update: A case study in fingerprints. *Pattern Recognition, 37*(7), 1533–1542. doi:10.1016/j.patcog.2003.11.012.

Vacca, J. R. (2007). *Biometric technologies and verification systems*. Burlington, MA: Butterworth-Heinemann.

Wang, C., & Gavrilova, M. (2006). Delaunay triangulation algorithm for fingerprint matching. In *Proceedings of ISVD*, (pp. 208-216). Banff, Canada: ISVD.

Wayman, J. L., Jain, A. K., Maltoni, D., & Maio, D. (2005). An introduction to biometric authentication systems. In Wayman, J. L., Jain, A. K., Maltoni, D., & Maio, D. (Eds.), *Biometric Systems: Technology, Design and Performance Evaluation* (pp. 1–20). London, UK: Springer-Verlag.

Wecker, L., Samavati, F., & Gavrilova, M. (2005). Iris synthesis: A multi-resolution approach. In *Proceedings of 3rd International Conference on Computer Graphics and Interactive Techniques in Australasia and South East Asia*, (pp. 121-125). IEEE.

Werner, C. (2008). *Biometrics: Trading privacy for security*. Retrieved from http://media.wiley.com/product_data/excerpt/26/07645250/0764525026.pdf

Wilson, C. (2010). *Vein pattern recognition: A privacy-enhancing biometric*. Boca Raton, FL: CRC Press. doi:10.1201/9781439821381.

Woodward, J. D. (1997). Biometrics: Privacy's foe or privacy's friend? *IEEE Proceeding, 85*(9), 1480-1492.

Woodward, J. D. Jr, Horn, C., Gatune, G., & Thomas, A. (2003). *Biometrics: A look at facial recognition*. Arlington, VA: Virginia State Crime Commission.

Yanushkevich, S., Gavrilova, M., Wang, P., & Srihari, S. (2007). *Image pattern recognition: Synthesis and analysis in biometrics*. New York, NY: World Scientific Publishers.

Chapter 3
Biometric Image Processing

ABSTRACT

In most biometric-based security systems, images of the associated biometric identifiers are used as the input to that system. This chapter discusses various image processing methods and algorithms commonly used for biometric pattern recognition. Efficient and reliable processing of images is essential to achieve good performance of biometric systems. Different appearance-based methods, such as eigenimage and fisherimage, and topological feature-based methods, such as Voronoi diagram-based recognition, are discussed in the context of face, ear, and fingerprint application frameworks. Utilizing cognitive intelligence and adaptive learning methods in both physical and behavioral biometrics are some emerging new directions of biometric pattern recognition. As such, neural networks, fuzzy logic, and cognitive architectures would play a more important role in biometric domain of research. The chapter concludes with discussion of the importance of context-based recognition for behavioral biometrics.

1. INTRODUCTION

In the previous chapter, an overview of biometric systems has been presented. For decades, many government and public establishments have used biometric authentication for access control. Today, the primary application of biometrics is shifting from the physical security, where the access to the specific locations is usually monitored using standard security identification mechanisms such as ID or token-based mechanisms in com-bination with fingerprint biometric, to remote security where the methods of crowd monitoring using video surveillance take advantage of gait biometric, for example. The popularity of such approaches has increased dramatically as the new technological devices are coming on the market every week, capability to process massive amount of data is doubling every few months, and the biometric algorithm development by the leading IT companies and the universities research centres is tripled in the last years.

DOI: 10.4018/978-1-4666-3646-0.ch003

As the demand for the development of more precise and reliable ways of person identification is ever pressing, the combination of biometrics and pattern analysis methods is gaining popularity as it undoubtedly increases the accuracy of the results and thus the level of security protection. Driven by the above motivations, the rest of this chapter is devoted to two main directions in intelligent biometric data processing. The first group of methods are appearance-based approaches, well known in biometric domain (Jain, Flynn, & Ross, 2007). The second group of methods are topology-based information driven methods (Yanushkevich, Gavrilova, Wang, & Srihari, 2007), with emphasis given to extracting metadata from biometric samples in order to simplify processing, reduce storage and increase accuracy, thus making approach to biometric processing more efficient and more intelligent.

2. APPEARANCE-BASED IMAGE PROCESSING IN BIOMETRICS

From the gamut of research on biometric authentication, we observe that the overwhelming part of biometric data processing is realized by using image processing and pattern recognition methods and algorithms (Soledek, Shmerko, Phillips, Kukharevl, Rogers, & Yanushkevich, 1997). As the mainstream direction of biometric image processing, appearance-based methods extract biometric features from the row image by analyzing appearance of the whole image as an entity or a vector in a high-dimensional image space. Such factors as color scheme, orientation, background, luminance, saturation are being analyzed and processed either pixel by pixel or through projection on subspaces, such as in Principal Component Analysis (PCA) methods. As the most evident examples, we consider face, iris and ear biometrics in this context (Soledek, Shmerko, Phillips, Kukharevl, Rogers, & Yanushkevich, 1997).

To solve the biometric data processing problems, the following main methods are typically employed in literature: digitization, compression, enhancement, segmentation, feature measurement, image representation, image models, design methodology. They are summarized in Figure 1. While some of these methods are used during data pre-processing, and some during pattern recognition and matching, there is high potential of employing more intelligent techniques at all stages, with the goal of optimizing processing and increasing overall security system performance. Some of these approaches are overviewed in the subsequent sections devoted to individual biometrics.

2.1. Image Processing for Face Recognition

The face matcher in a security biometric system is usually used for face recognition. Its' main goal is to identify recognizable facial characteristics from images, to reduce the key features to digital codes, and to match them against known facial templates. The inputs to the matcher are the input image and the face images from a facial image database, and the output is a single matched face or a ranked list with the top-n matches, i.e. first n recognized match faces. The output is enough on its own to make a decision to grant or not access to a given resource or secure asset, and also is suitable for further fusion as part of rank-level multi-modal biometric system. This procedure will be described in details in chapter on Rank-Level Multimodal Biometric System Architecture.

In order to recognize faces, first features from the face images have to be extracted and selected to represent the properties of the data in the most effective way for future match computation in the feature space. The goal is to extract most important features to *differentiate* or *separate* individuals in the biometric facial space. The distance between such features is further computed using selected

Figure 1. Biometric image processing methods and algorithms

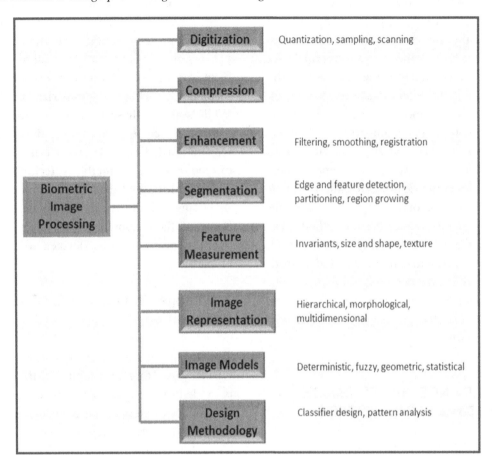

distance metric space (Euclidean, Minkowski, Lyapunov, Mahalanobis, Geodesic, etc.). Many approaches to selecting and extracting both the distance metrics and the features have been suggested in the pattern recognition literature. The choice is often dictated by the specifics of the problem itself. Thus, the Euclidean metric is often used to solve the geometric problems in Cartesian space, while Mahalanobis metric is used for statistical analysis of variance.

Among various approaches to face recognition, some of the most popular and effective are the appearance-based approaches. Principal Component Analysis (PCA) and Linear Discriminant Analysis (LDA) are two examples of such methods working on principle of dimensionality reduction

and feature extraction.(Belhumeur, Hespanha, & Kriegman, 1997; Bartlett, Movellan, & Sejnowski, 2002; Lu, Plataniotis, & Venetsanopoulos, 2003). Two state-of-the-art face recognition methods, Eigenfaces (Turk & Pentland, 1991) and Fisherfaces (Belhumeur, Hespanha, & Kriegman, 1997), built on these two techniques, respectively, have been proved to be very successful. In PCA, object classes that are neighbors in the output space are often weighted in the input space to reduce potential misclassification. According to (Bartlett, Movellan, & Sejnowski, 2002), the PCA could be either used over raw face image to extract the features or over the eigenface to extract the discriminant Eigenfeatures. From the gamut of literature which followed the invent of

PCA, it has been shown that this technique is very sensitive to image conditions such as background noise, image shift, occlusion of objects, scaling of the image, illumination changes, and others. Thus, though PCA method is still one of the most popular techniques for face recognition, there has been a significant interest in method improvement.

Numerous variations of PCA combining feature representation methods that utilize the strengths of different realizations have been proposed. LDA (Linear Discriminant Analysis) (Belhumeur, Hespanha, & Kriegman, 1997), Kernel PCA (Kim, Jung, & Kim, 2002), and Generalized Discriminant Analysis (GDA) using a kernel approach (Baudat & Anouar, 2002) all were suggested for specific applications. One of the popular approaches is to use Fisherimage, which is a combination of PCA and LDA for facial image recognition (Belhumeur, Hespanha, & Kriegman, 1997). The method obtains a subspace projection matrix. Eigenimage method attempts to maximise the scatter matrix of the training images in image space, while fisherimage method attempts to maximise the between class scatter matrix (also called extra-personal) while at the same time minimising the within class scatter (intra-personal) (Figure 2). In fisherimage method, images of the same face class are projected closer together, while images of difference faces end up further apart. Further, fisherimage method has

several other advantages. This method is more robust against noise and occlusion, against illumination, scaling and orientation; and against varied facial expressions, facial hair, glasses, and makeup. In addition, fisherimage method can handle high resolution or low resolution images efficiently, and can provide faster recognition with low computational cost. The implementation for this method is summarized below, based on the formulas presented in Turk and Pentland (1991), Belhumeur, Hespanha, and Kriegman (1997), and Zhang (2004).

According to Turk and Pentland (1991), one must first initialize the system with a set of training set of face image vectors containing multiple images of each subject as:

$$\text{Training set} = \left\{ \underbrace{\Gamma_1 \Gamma_2 \Gamma_3 \Gamma_4 \Gamma_5}_{X_1} \underbrace{\Gamma_6 \Gamma_7 \Gamma_8 \Gamma_9 \Gamma_{10}}_{X_2} \underbrace{\Gamma_{16} \Gamma_{17}}_{X_4} \ldots\ldots\ldots \underbrace{\ldots \Gamma_N}_{X_c} \right\} \quad (3.1)$$

where Γ_i is a face image vector, N is the total number of images and each image belongs to one of c classes $\{X_1, X_2, ..., X_C\}$ where C is the number of subject in the database.

The face image vector Γ can be obtained by reconstructing the original face image by adding each column one after another. Thus, a face image

Figure 2. Examples of between class and within class scatters

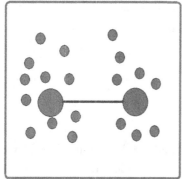

Between class scatter

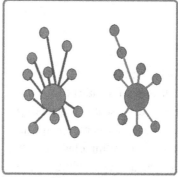

Within class scatter

represented by $(N_x \; X \; N_y)$ pixels can be reconstructed into an image vector Γ of size $(P \; X \; 1)$, where P is equal to $(N_x * N_y)$.

Then, the between class scatter matrix (S_B) and the within class scatter matrix (S_W) can be defined according to (Turk & Pentland, 1991) by the following equations:

$$S_B = \sum_{i=1}^{C} |X_i| (\Psi_i - \Psi)(\Psi_i - \Psi)^T$$

$$S_B = \sum_{i=1}^{C} |X_i| (\Psi_i - \Psi)(\Psi_i - \Psi)^T \qquad (3.2)$$

$$S_W = \sum_{i=1}^{C} S_i \qquad (3.3)$$

where, $\Psi = \dfrac{1}{N} \sum_{i=1}^{N} \Gamma_i$ is the arithmetic average of all the training image vectors in the database at each pixel points and its size is $(P \; X \; 1)$. $\Psi_i = \dfrac{1}{|X_i|} \sum_{\Gamma_i \varepsilon X_i}^{N} \Gamma_i$ is the average image of class Xi at each pixel points and $|X_i|$ is the number of samples in class X_i and its size is $(P \; X \; 1)$. The mean or average face images of each class are necessary for the calculation of each face class's inner variation. S_i is the scatter of class i which is defined as (Turk & Pentland, 1991):

$$S_i = \sum_{\Gamma_i \varepsilon X_i} \left(\Gamma_i - \Psi_i\right)\left(\Gamma_i - \Psi_i\right)^T \qquad (3.4)$$

The size of the between class scatter matrix (S_B) and the within class scatter matrix (S_W) are both $(P \, X \, P)$. The between class scatter matrix (S_B) represents the scatter of each class mean around the overall mean vector. The within class scatter matrix (S_W) represents the average scatter of the image vectors of different individuals around their respective class means.

After defining between class scatter matrix (S_B) and the within class scatter matrix (S_W), one can define the total scatter matrix S_T of the training set as (Belhumeur, Hespanha, & Kriegman, 1997):

$$S_T = \sum_{i=1}^{N} (\Gamma_i - \Psi)(\Gamma_i - \Psi)^T \qquad (3.5)$$

The objective of using Fisher's Linear Discriminant is to classify the face image vectors. A commonly used method to do so is to maximize the ratio of the between class scatter matrix of the projected data to the within-class scatter matrix of the projected data. Thus, an optimal projection W which maximizes between-class scatter and minimizes within-class scatter can be found by the following equation (Belhumeur, Hespanha, & Kriegman, 1997):

$$W = \max\left(J\left(T\right)\right) \qquad (3.6)$$

where $J(T)$ is the discriminant power which can be obtained by the following equation:

$$J_T = \frac{\left| T^T \cdot S_B \cdot T \right|}{\left| T^T \cdot S_W \cdot T \right|} \qquad (3.7)$$

In the above equations, S_B and S_W are the between class scatter matrix and within class scatter matrix, respectively. Hence, the optimal projection matrix W can be re-written as:

$$W = \max\left(J\left(T\right)\right) = \max \left. \frac{\left| T^T \cdot S_B \cdot T \right|}{\left| T^T \cdot S_W \cdot T \right|} \right|_{T=W} \qquad (3.8)$$

and can be obtained by solving the generalized eigenvalue problem:

$$S_B W = S_W W \lambda_W \qquad (3.9)$$

with λ is the eigenvalue of the corresponding eigenvector (Belhumeur, Hespanha, & Kriegman, 1997).

From the generalized eigenvalue equation, no more than c-1 of the eigenvalues are nonzero, and only the eigenvectors coming out with these nonzero eigenvalues can be used in forming the W matrix. Once the W matrix is constructed, it is used as the projection matrix. The training image vectors are projected to the classification space by the dot product of the optimum projection W and the image vector as follows (Belhumeur, Hespanha, & Kriegman, 1997):

$$g\left(\Phi_i\right) = W^T \cdot \Phi_i \qquad (3.10)$$

where Φ_i is the mean subtracted image:

$$\Phi_i = \Gamma_i - \Psi_i \qquad (3.11)$$

This projection matrix is of the size ($(c-1)$ x 1). And its components can be viewed as images, referred to as fisherimages (Belhumeur, Hespanha, & Kriegman, 1997).

After enrolling the face images, one can organize the recognition output which lists first n best matches. This can be used on its own or as an input to rank level fusion module for further decision making in multi-modal system. To achieve this, the following tasks can be performed:

Step 1: Project the test face image into the fisherspace, and measure the distance between the unknown face image's position in the fisherspace and all the known face image's positions in the fisherspace. The projection of the test image vector to the classification space is done in the same manner:

Classification space projection,

$$g\left(\Phi_T\right) = W^T \cdot \Phi_T \qquad (3.12)$$

which is of the size ($(c-1)$ x 1).

One of the simple ways to compute the distance between the projections is to calculate it by the Euclidean distance between the training and test classification space projections.

$$d_{Ti} = \left\| g\left(\Phi_T\right) - g\left(\Phi_i\right) \right\| \qquad (3.13)$$

Step 2: Select the image closest to the unknown image in the fisherspace.
Step 3: Repeat Step 2 (without considering the match image obtained through Step 2) until all n best match images obtained. *Stop.*

Figure 3 presents general flowchart for fisherface generation process and Figure 4 presents sample fisherface generated in a biometric system.

It must be noted that traditional face recognition algorithm described above has recently being challenged by emerging intelligent image processing applications. In that group, face recognition techniques based on elastic graph matching, neural network learner and Support Vector Machines (SVMs) have proven to be highly promising (Gavrilova & Ahmadian, 2011; Kim, Jung, & Kim, 2002; Baudat & Anouar, 2002; Rowley, Baluja, & Kanade, 1998; Phillips, 1998). These methods not only identifying common and most distinguishing patterns based on facial biometrics, but also are highly adaptive to changing environment, data quality, and application domains.

Among those methods, approach based on neural-network learner and reducing complexity of input data vectors received a lot of attention (Gavrilova & Ahmadian, 2011). This novel direction for facial recognition can potentially yield

Figure 3. Sample input and acquired signals for a biometric system (image sources: CASIA and Google)

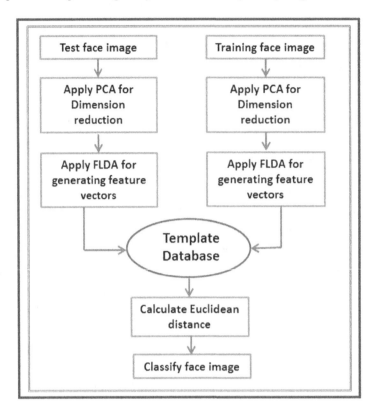

Figure 4. Sample fisherfaces generated from a face database

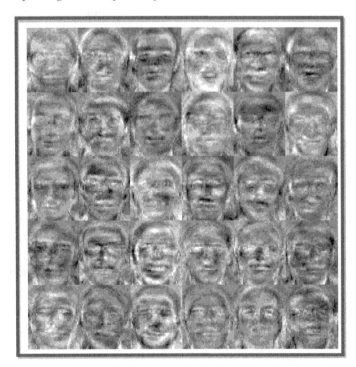

a significant improvement is overcoming high-complexity and high-dimensionality of biometric data. The goal is to transform data from a high-dimensional space into a lower-dimensional one without the loss of the information. Normally, the lower dimension maximizes the variance of data. High-dimensionality of data is a typical problem in biometric recognition systems when multiple features from the training samples are used. The complexity of designing algorithms for the recognition purpose grows significantly as the number of dimensions grows.

A common set of methods to reduce the dimensionality of space is the clustering approach. In clustering, the elements of the set are grouped according to their similarity by some similarity measures. Usually clustering is used to design a set of boundaries to better understand the data (based on structured data). Other usage of clustering involves indexing and data compression. By creating a meaningful subspace of the original space and then providing this reduced vector space to neural network or evolutionary method learner, one can achieve both higher accuracy results and better system sustainability in the presence of low quality data. This approach will be described in more details in subsequent chapters.

2.2. Iris Recognition Algorithms

The iris processing is similar in a way to face recognition pattern-matching as it is based on appearance-based feature detection method, tailored specifically to take in the account the iris physiological structure. This method is described below.

Iris is a plainly visible ring that surrounds the pupil of an eye (Vacca, 2007). It is a muscular structure that controls the amount of light entering an eye, with intricate details that can be measured, such as, striations, pita, and furrows (Vacca, 2007). An iris recognition system first creates the measurable features of the iris. These features are then stores and later compared with new algorithms

of irises presented to a capturing device for either identification or verification purpose.

One of the iris recognition methods (Daugman, 1993) uses Hamming distance (Hamming, 1950) for iris matching after the iris image is pre-processed and encoded with Hough transform (Hough, 1962) and 2-D Gabor wavelet (Gabor, 2012). At first, one needs to localize the iris part of the eye image (from inside the limbus (outer boundary) and outside the pupil (inner boundary)) using an automatic segmentation algorithm based on Hough transform (Hough, 1962). The *Hough transform* method is a general technique for identifying the locations and orientations of certain types of features in a digital image and has a several advantages (Hough, 1962). This method is simple, easy to implement, handles missing and occluded data well, and can be adapted to many types of forms, not just lines.

For segmenting the iris from the eye image, Hough transform (Hough, 1962) method is used. For iris region extraction, a circular Hough transform method is employed in which circular iris edge points are extracted through a voting mechanism in the Hough space. Two edge detected images of the original eye images – one with the horizontal gradients and the other with the vertical gradients, are generated for efficiently isolation of the iris boundary. Figure 5 illustrates this process.

After localizing the pupil and iris in the eye image, we can store the radius and the x and y centre coordinates for both circles (pupil and iris). Then, one can isolate the eyelids by fitting a line to the upper and lower eyelid using the linear Hough transform (Hough, 1962). Another horizontal line, which intersects the first line, is then used to isolate the eyelid regions.

The iris region is then transformed into polar coordinates system to facilitate the feature extraction process. For this, first the portion of the pupil from the conversion process has been excluded because it has no biological characteristics. Next, the transformation process is applied to bring characteristic features of different irises to the

Figure 5. The eye image from CASIA database and corresponding horizontal and vertical edge maps

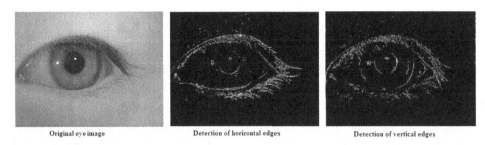

| Original eye image | Detection of horizontal edges | Detection of vertical edges |

same spatial position. One can use the rubber sheet model (Daugman, 1993) to remap each point within the iris region to a pair of polar co-ordinates ((r, θ), where r lies in the interval $[0,1]$ and θ is the angular variable, cyclic over $[0,2\pi]$. This remapping of the iris region can be modeled as (Daugman, 1993):

$$I(x(r, \theta), y(r, \theta)) \rightarrow I(r, \theta) \quad (3.14)$$

with $x(r, \theta) = (1-r)\, x_p(\theta) + rx_i(\theta)$

$$y(r, \theta) = (1-r)\, y_p(\theta) + ry_i(\theta) \quad (3.15)$$

where $I(x,y)$ is the iris region image, (x,y) are the original Cartesian coordinates, (r,θ) are the corresponding normalized polar coordinates, and (x_p, y_p) and (x_i, y_i) are the coordinates of the pupil and iris boundaries along the θ direction. (Daugman, 1993).

Then, one can encode the normalized iris pattern into an iris code through a process of demodulation (introduced in Daugman, 1993) that extracts phase sequences using a 2-D Gabor wavelets:

$$h_{\{Re,Im\}} = \mathrm{sgn}_{\{Re,Im\}} \int_\rho \int_\varphi I(\rho, \varphi) e^{-iw(r_0-\rho)^2/\alpha^2}$$
$$\times e^{-(\theta_0-\varphi)} e^{-(\theta_0-\varphi)^2/\beta^2} \rho d\rho d\varphi$$
$$(3.16)$$

where $h_{\{Re,Im\}}$ is a complex-valued bit whose real and imaginary parts are either 1 or 0 (sign) de-

pending on the sign of the 2-D integral; $I(\rho,\varphi)$ is the raw iris image, α and β are the multi scale 2-D wavelet size parameters, ω is wavelet frequency and (r_0, θ_0) represent the coordinates of each region of iris for which the phasor bits $h_{\{Re,Im\}}$ are computed. The overall iris code generation process is illustrated in Figure 6.

The next step is comparing two code-words to find out if they represent the same person or not. Once again, Hamming distance method is suitable for this purpose. The method is based on the idea that the greater the Hamming distance between two iris feature vectors, the greater the difference between them. The Hamming Distance (HD) between two Boolean iris vectors is defined as follows (Daugman, 1993):

$$HD = \frac{\left\| C_A \otimes C_B \cap M_A \cap M_B \right\|}{\left\| M_A \cap M_B \right\|} \quad (3.17)$$

where C_A and C_B are the coefficients of two iris images and M_A and M_B are the mask image of two iris images. The $\otimes$ is the XOR operator which shows difference between a corresponding pair of bits, and $\cap$ is the AND operator which shows that the compared bits are both have not been impacted noise. d. As with any method, even when the distance points to a perfect match, the two irises might not be identical due to noise or errors in processing or distance computation. Nevertheless, iris recognition based on hamming distances considered to be one of the strongest

Figure 6. Iris code generation process

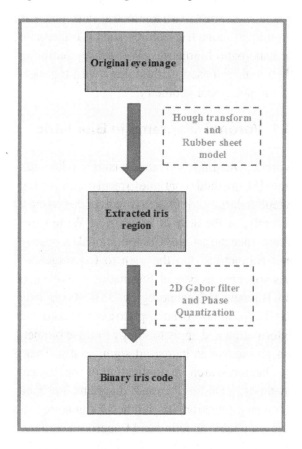

2.3. Appearance-Based Ear Recognition

The ear is a biometric, which has not received enough attention in commercial applications; mostly due to existence of much more commonly accepted authentication biometrics (such as face, fingerprint and iris). However, certain features of ear biometric, such as being image-based (similarly to face), permanent over time, fairly unique for each individual and available to obtain by traditional image capturing methods (i.e. using camera or extracted from facial profile images), make it worth mentioning.

Some of the well known examples of ear images can be found in IIT Delhi Ear Image Database, collected at IIT Delhi, New Delhi, India (IIT Delhi, 2012). This database has been assembled at IIT Delhi campus from October 2006 to June 2007. All images were acquired from a distance (touchless method) using simple imaging setup in the indoor environment. The database of almost five hundred images has been sequentially numbered for every user with an integer identification number.

Some of the popular ear recognition approaches look at ear furrows and valleys and applies minimization process to find common ear features (similarly to minutiae matching in fingerprints) (Jain, Flynn, & Ross, 2007). However, an appearance-based method based on eigenfaces or feasherfaces, described above, can be just as efficient for ear recognition. Thus, after normalizing the images, the feature extraction is performed. There are several techniques to retrieve features from ear images, based on methods such as geometrical distance metrics and Haar wavelet transform. First, a sample mean images of ear database are obtained (See Figure 7). Next, we can use dimensionality-reduction methods directly over image pixels. The known methods which can extract meaningful features from such information are Maximum Variance Unfolding (MVU),

biometrics up to date. The issues with their practical application in real biometric security systems are in data sampling, availability and cost of processing.

As with the face recognition method, the resulting top-n matches are reported as the outcome. These matches can also be considered for information fusion when iris processing module is part of multi-modal biometric system. Then, the templates will be sorted according to the Hamming distance in the ascending order and the top-n templates are used as input for the rank level fusion.

Laplacian Eigenmaps, Multidimensional Scaling (MS), or Subspace Clustering (SC) (Gavrilova & Ahmadian, 2011).

As a result of this procedure, a small number of most prominent ear features is chosen. Next, the proper number of eigenvectors for the recognition purpose is determined. In order to do that, one may consider first *k* eigenvectors (sort based on the corresponding eigenvalues) and pick the k in such a way that the correlation between the values is low, which makes these eigenvectors important for the recognition phase. To illustrate the concept, the correlation rate of different eigenvectors obtained by Subspace Clustering dimensionality-reduction method from raw data is shown in Figure 8. As it can be seen, the initial 3D data is first reconstructed based on first two Eigenvalues, next 2D projection matrix is clustered, and finally the 3d initial data set is clustered.

3. TOPOLOGY-BASED INTELLIGENT PATTERN ANALYSIS IN BIOMETRICS

The goal of any intelligent processing is to minimize overhead associated with performing computations while at the same time to maximize an output. The same principle governs behavior of most public and commercial organizations—to achieve high production by resource and processes optimization. While appearance-based methods excel in capturing even subtle features in the multitude of high-dimensional data, sometimes generalizing the results and noting common patterns leads to process optimization without sacrificing the security system performance.

This section presents topology-based methods, which work best with biometric data that has prominent geometric features, such as fingerprint or hand/palm biometrics. We start by outlining the topology-based methodology with the roots in computational geometry.

3.1. Voronoi Diagrams in Biometric

Voronoi Diagram (VD) and Delaunay Triangulation (DT) methods continue to receive compelling attention in the various areas of research, and most recently, in the area of *biometrics*. Within two years since the article "Computational Geometry and Biometrics: On the Path to Convergence" has appeared as part of International Workshop on Biometrics Technologies 2004 (Gavrilova, 2004), the number of attempts to extract geometric information and apply topology to solve biometric problems has increased significantly. There has been research on application of topological methods, including Voronoi diagrams, for hand geometry detection, iris synthesis, signature recognition, face modeling and fingerprint recognition (Bebis, Deaconu, & Georiopoulous, 1999; Wang & Gavrilova, 2004; Wecker, Samavati, & Gavrilova, 2005; Xiao & Yan, 2002). While the methodology is rooted in computational geometry domain, many questions on what is the best way to utilize this data structure, which topological information to use and in which context, how to make implementation decisions, and whether the performance is comparable with other biometric feature extraction techniques still remain. The use of geometric information in application to biometrics is a new area of research. The Voronoi diagram

Figure 7. IIT Delhi ear mean samples (IIT Delhi, 2012)

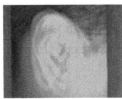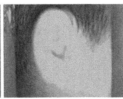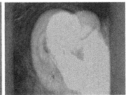

Figure 8. The eigenvectors in subspace clustering

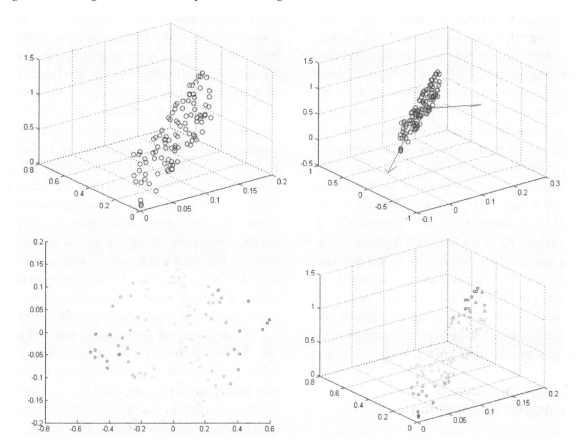

and Delaunay triangulation are two fundamental geometric structures that can be used to describe the topological structure of the fingerprint, which is considered to be the most consistent information for fingerprint matching purposes. Here, we first introduce some basic definitions.

A Voronoi region associated with a feature is a set of points that are closer to that feature than to any other feature. Given a set S of points p_1, $p_2, \ldots, p_n$, the Voronoi diagram decomposes the space into regions around each point such that all points in the region around p_i are closer to p_i than to any other point (Preparata & Shamos, 1985). Let $V(S)$ be a Voronoi diagram of a planar point set S. Consider the straight-line dual $D(S)$ of $V(S)$, that is, the graph obtained by adding an edge between each pair of points in S whose Voronoi regions share an edge. The dual of an edge in $V(S)$ is an edge in $D(S)$. $D(S)$ is a triangulation of the original point set, and is called the Delaunay triangulation (Preparata & Shamos, 1985). There are some compelling reasons why the Delaunay triangulation is a naturally suitable data structure for the alignment of minutiae set. The Delaunay triangulation is uniquely identified by the set of points. Inserting or removing a new point typically has only local effect on the triangulation, which means the algorithm is tolerant to some imprecision of minutiae extraction technique. The Delaunay edges are on average shorter than edges connecting two randomly chosen minutiae, as the Delaunay triangulation is regular triangulation. Considering the elastic deformation of fingerprint, when the distance between two points in template image is increases, it becomes more challenging to match point pairs in the corresponding images.

Introducing deformation tolerant method based on topological information seems as a logical and promising way to deal with the problem. Finally, use of additional topological information extracted from ridge geometry or based on singular points in addition to minutia set allows to significantly improve matching accuracy at the negligibly small computational overhead. The outlined above ideas are described in the next subsection and were fully implemented in the fingerprint recognition software developed at BTLab (Wang, Gavrilova, Luo, & Rokne, 2006). The experimental results based on comparison of this system with the other traditional as well as recent methods clearly demonstrate the advantages of the Delaunay-based techniques. While detailed numerical results have appeared in (Wang, Gavrilova, Luo, & Rokne, 2006), the basic idea of using both global and local geometrical features for fingerprint recognition system is outlined in the following sections.

3.2. Topology-Based Fingerprint Recognition

As agreed by the experts, the main methodologies for fingerprint matching are typically:

- Correlation-based matching (Ratha, Karu, Chen, & Jain, 1996)
- Minutiae-based matching (Jain, Hong, & Bolle, 1997)
- Ridge-feature-based matching (Jiang & Yau, 2000)

In correlation-based matching, two fingerprint images are overlapped and the relative correlation between corresponding pixels is computed (Ratha, Karu, Chen, & Jain, 1996). During minutiae-based matching, the set of minutiae are extracted from the two fingerprints and then compared in the two dimensional plane (Jain, Hong, & Bolle, 1997). Ridge-feature-based matching is based on ridge geometry including their orientation (Jiang

& Yau, 2000). Minutiae-matching has been one of the most popular approaches that yields very good matching results. On the perpetual quest for perfection, a number of techniques were devised for reducing FAR (False Acceptance Rate) and FRR (False Rejection Rate); computational geometry being one of such techniques. Thus, Voronoi diagrams were utilized for face partitioning onto segments and facial feature extraction in (Xiao & Yan, 2002). Bebis et al. (Bebis, Deaconu, & Georiopoulous, 1999) used the Delaunay triangle as the comparing index in fingerprint matching. The method works under assumption that at least one corresponding triangle pair can be found between the input and template fingerprint images. Unfortunately, in real world situation this assumption might not hold due to low quality of fingerprint images, unsatisfactory performance of the feature extraction algorithm or distorted image. Another problem in fingerprint matching is noticeable when the distance between two points in template image is increases. It becomes much more challenging to match point pairs in the corresponding images. For instance, (Kovacs-Vajna, 2000) have shown that local deformation of less than 10% can cause global deformation reaching 45% in edge length. In order to deal with those shortcomings, the method combining both global and local fingerprint matching and applying radial functions for deformation modeling has been developed in BTLab at the University of Calgary (Wang & Gavrilova, 2005; Wang, Gavrilova, Luo, & Rokne, 2006). Its main ideas are briefly described below.

The method utilizes geometric invariant features of Delaunay triangulation for minutiae matching and singular-point comparison in the fingerprints. The method was implemented as part of the global fingerprint recognition system and is shown to be sustainable under presence of elastic deformations. The minutiae-matching algorithm, in which context the Delaunay triangulation approach is utilized, is based on classic fingerprint matching technique. Using Delaunay triangulation brings unique challenges and advantages. First of all, Delaunay edges rather than minutiae or whole minutiae triangles are selected as matching index which provides an easier way to compare two fingerprints. Secondly, the method is combined with deformation model, which helps to preserve consistency of the results under elastic finger deformations. Third, to improve matching performance, features based on spatial relationship and geometric attribute of ridges are introduced, and further combined with information from both singular point sets and minutiae sets to increase matching precision.

The purpose of fingerprint identification is to determine whether two fingerprints are from the same finger or not. In order to do this, the input fingerprint needs to be aligned with the template fingerprint represented by its minutia pattern. The following rigid transformation can be performed:

$$F_{s,\Delta\theta,\Delta x,\Delta y} \begin{pmatrix} x_{templ} \\ y_{templ} \end{pmatrix}$$
$$= s \begin{pmatrix} \cos\Delta\theta & -\sin\Delta\theta \\ \sin\Delta\theta & \cos\Delta\theta \end{pmatrix} \begin{pmatrix} x_{input} \\ y_{input} \end{pmatrix} + \begin{pmatrix} \Delta x \\ \Delta y \end{pmatrix} \quad (3.18)$$

where $(s, \Delta\theta, \Delta x, \Delta y)$ represent a set of rigid transformation parameters (scale, rotation, translation). Under a simple affine transformation, a point can be transformed to its corresponding point after rotating $\Delta\theta$ and translating $(\Delta x, \Delta y)$.

The algorithm has three distinctive stages. During the first stage, identification of feature patterns utilizing Delaunay triangulation takes place. During the second stage, Radial Basis Function (RBF) is applied to model the finger deformation and align images. Finally, global matching is employed to compute the combined matching score, using additional topological information extracted from ridge geometry (See Figure 9).

Let $Q = ((x_1^Q, y_1^Q, \theta_1^Q, t_1^Q)...(x_n^Q, y_n^Q, \theta_n^Q, t_n^Q))$ denote the set of n minutiae in the input image ((x,y): location of minutiae; θ : orientation field of minutiae; t: minutiae type, end or bifurcation;) and $P = ((x_1^P, y_1^P, \theta_1^P, t_1^P)...(x_m^P, y_m^P, \theta_m^P, t_m^P))$ denote the set of m minutiae in template image.

Table 1 shows topological features that we can use for both local and global matching of fingerprints. In the above table, *Length* is the length of edge; θ_1 is the angle between the edge and the orientation field at the first minutiae point; *Type*$_1$ denotes minutiae type of the first minutiae; *Ridge count* is the number of ridges that these two minutiae points cross. Using triangle edge as comparing index has many advantages. For local matching, once we compute the Delaunay triangulation of minutiae sets Q and P, we use triangle edges for match determination as comparing index. To compare two edges, *Length*, θ_1, θ_2, *Type*$_1$, *Type*$_2$, *Ridgecount* parameters are used. Note that these parameters are invariant of the translation and rotation (See Table 1). Conditions for determining whether two edges match identified as the set of linear inequalities depend on the above parameters and the specified thresholds (usually derived empirically depending on image size and quality). A sample set of such conditions can be found in (Wang & Gavrilova, 2004). If the threshold is selected successfully, transformation $(\Delta\theta, \Delta x, \Delta y)$ is used to align the input and template images. Fingerprint matching using Delaunay triangulation edge is then realized as follows. If one edge from an input image matches two edges from the template image, we need to consider the triangulation to which this triangle edge belongs to and compare the triangle pair. For a certain range of translation and rotation dispersion, we detect the peak in the transformation space, and record transformations that are neighbors of the peak in the transformation space. Note that those recorded transformations are close to

Figure 9. Flow chart of generic fingerprint identification system (Wang, Gavrilova, Luo, & Rokne, 2006)

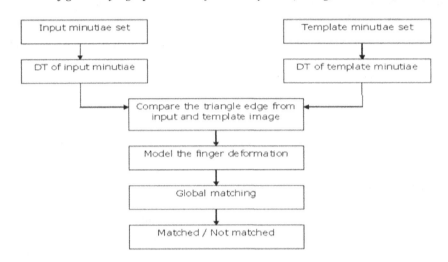

Table 1. Topological properties of Delaunay Triangulation used for fingerprint image matching (Y – dependent on fingerprint transformations, N – independent on them)

Feature	Fields					
Minutiae Point	*x (Y)*	*y (Y)*	*θ (Y)*	*Type (N)*		
Triangle Edge	*Length (N)*	*θ₁ (N)*	*θ₂ (N)*	*Type₁ (N)*	*Type₂ (N)*	*Ridge count (N)*

each other, but not identical. Figure 10 shows the successfully matched Delaunay triangle edge pairs.

The deformation problem is rooted in the inherent flexibility of the finger. Pressing a finger against a flat surface introduces distortions which need to be accounted for. A good fingerprint identification system should always compensate for these deformations. In Wang and Gavrilova (2005), we proposed a framework aimed at quantifying and modeling the local, regional and global deformation of the fingerprint. The method is based on the use of Radial Basic Functions (RBF), which represent a practical solution to the problem of modeling of a deformable behavior (Cappelli, Maio, & Maltoni, 2001). For the fingerprint-matching algorithm, deformation problem can be described as knowing the consistent transformations of specific *control* points from the minutiae set, and knowing how to interpolate

the transformation of other minutiae which are not control points. Given the coordinates of a set of corresponding points (control points) in two images: $\{(x_i, y_i), (u_i, v_i) : i = 1, ..., n\}$, determine function $f(x, y)$ with components $f_x(x, y)$ and $f_y(x, y)$ such that

$$u_i = f_x(x_i, y_i),$$
$$v_i = f_y(x_i, y_i), \quad i = 1, ..., n. \quad (3.19)$$

Rigid transformation can be decomposed into a translation and a rotation (See Equation(3.18)). In 2D, a simple affine transformation can represent rigid transformation as:

$$f_k(\vec{x}) = a_{1k} + a_{2k}x + a_{3k}y \quad k = 1, 2 \quad (3.20)$$

where $\vec{x} = (x, y)$.

Figure 10. Successfully matched Delaunay triangle edge pairs

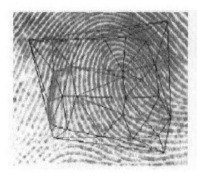 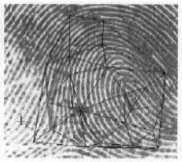

There are three unknown coefficients in this rigid transformation which can be solved by known methods for finding unknown in linear system. Figure 11 shows the alignment of two images (a) and (b) when we represent the transformation of input image as a rigid transformation. The maximum number of matching minutiae pairs between input image and template image is 6 (see (c)). Circles denote minutiae of the input image after transformation, while squares denote minutiae of the template image. Knowing transformation of minutiae (control points) in the input image, we apply the RBF (Cappelli, Maio, & Maltoni, 2001) to model the non-rigid deformation which is shown in (d). The resulting number of matching minutiae pairs is thus ten.

After local matching procedure was performed and deformation model applied to minutiae set, the numbers of paired and matched minutiae can be obtained globally. If two minutiae fall into the same tolerance box after identification, they are defined as paired. The tolerance threshold can be determined from experimentation, and is related to real system operational conditions.

Sometimes, additional features may be used in conjunction with the minutiae to increase the system accuracy and robustness. Additional features based on the spatial relationship and geometry attributes of ridge lines can be used in addition to edges and triangles, thus creating even more intelligent and reliable fingerprint recognition method. Using other distance functions to determine the threshold can be another possible direction of future research.

4. MODEL-BASED BEHAVIORAL BIOMETRICS

Identifying patterns in behavioral biometrics, in general, is a slightly different and somewhat more complex problem than identifying features in physiological biometrics. Examples of behavioral biometrics include signature, voice, gait, and typing patterns. Due to temporal dynamic features associated with each biometric (samples must be observed over period of time for best matching results), these problems are often treated in a class of signal-processing methods. In a nutshell, the task and the overall biometric system architecture remain the same, however upon closer examination; some very specialized methods taking advantage of unique continuous nature of those biometrics have been developed (Wayman, Jain, Maltoni, & Maio, 2006). We will illustrate the notion on example of gait analysis. Gait analysis deals with analyzing the patterns of walking movement. The fundamental work in gait analysis attributed to Johansson (1973) who showed that people can quickly recognize the motion of walking only by observing the moving patterns of lights attached to the moving body. Inspired by that work, Cutting and Kozlowski (1977) demonstrated that the same array of point lights can be used to recognize subjects even if

Figure 11. Benefits of rigid transformation in fingerprint matching

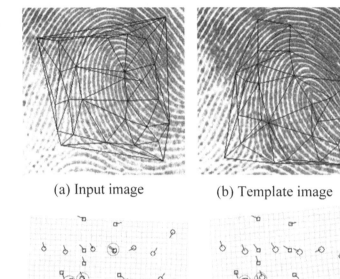

(a) Input image   (b) Template image

(c) Rigid transformation   (d) Non-rigid transformation

they happen to have similar height, width and body shapes. Considering the wide variety of potential applications for gait analysis, these studies open the door to further active research in this field. Although gait analysis is most well known for its application in access control, surveillance and activity monitoring, it can also be used in sports training to analyze the athlete's movements and give suggestions for improvement. Medical sciences can also take advantage of gait analysis techniques in diagnosing and treating certain patients with walking disorders.

The focus of gait analysis is the same as for any other biometric: identifying people from the way they walk or gait recognition. Gait recognition has recently attracted more attention due to a set of unique and interesting properties. This trait is unobtrusive meaning that the attention or

cooperation of the subject is not needed for collecting the data. Unlike many other biometrics, no specially designed hardware is needed. A surveillance camera is often sufficient for data acquisition and data collection can be both overt or covert, meaning to take place with or without knowledge of the subject. This trait is remotely observable and the subject does not even need to be close to the camera (Wang, 2005). These two properties make the data collection process more convenient compared to other biometrics. In addition, imitating the walking style of another person can be quite difficult. It is not easy to conceal the way you walk (Liu & Zheng, 2007) and it is possible to recognize patterns when person is trying to walk differently. Finally, gait recognition techniques usually don't need high resolution video sequences (Wang, She, Nahavandi, & Kouzani,

2010) and since they work on binary silhouettes they are not extremely sensitive to illumination changes and can even be used at night (Cuntoor, Kale, & Chellappa, 2003).

However, as any biometric, gait recognition suffers from limitations and challenges. Age, mood, illness, fatigue, drug, or alcohol consumption can affect the walking style of a person. Types of shoes, surface, and visibility can impact a person's walking style. In addition, any factor that can change the appearance of the person, such as wearing a hat, carrying a suitcase, having loose clothing, can adversely affect the performance of the gait recognition techniques. While temporal signal-processing and model-based approaches can capture many characteristics essential for successful recognition, sometimes this is simply not good enough. In summary, the main drawback of using gait for individual identification compared to other biometrics is its wide variability per subject. The novel idea thus is to augment the imperfect biometric such as Gait recognition by using supplementary context-based analysis which allows to reveal even stronger connections from data samples to the *social context* in which those samples have been obtained. This exciting area of research will be presented in Chapter 10.

5. SUMMARY

In this chapter, different image processing methods and algorithms that are popular in biometric data processing has been presented. In the case of the most of the biometric identifiers used today, image of that identifier is mainly the input to the biometric system. Thus, the processing of the biometric images is very essential for efficient and reliable performance of the biometric system. Usually, the main methods which are used for biometric image processing are digitization, compression, enhancement, segmentation, feature measurement, image representation, image models and design methodology. The feature extraction methods have been classified as appearance-based

and topological feature-based, and illustrated on example of different biometrics. First eigenvector and fisherface methods for face and ear biometric have been presented. Then there has been a discussion on original Voronoi diagram based methodology for feature matching in fingerprint images. The chapter is concluded with discussion of importance of context-based recognition for behavioral biometrics.

As with any biometric research, there are some emerging new directions, which authors believe would become mainstream in the near future. One of those is utilizing cognitive intelligence and adaptive learning methods in both physical and behavioral biometrics. For instance, the matching method and features extracted from row data would be dependent on the application environment, specific system operational requirements, and the data set. Training with specific biometric data would be beneficial to increase recognition rates and to minimize the impact of noise and imperfection in data capturing or processing. Moreover, the system design instead of being traditional (sequential/parallel) architecture, which most of the biometric systems existing today exhibit, would be resembling more and more cognitive processes of the human brain and decisions made by real humans. As such, neural networks, fuzzy logic, and cognitive architectures would play more and more important role in biometric domain of research.

We are now ready to consider another recent direction of biometric research, focused on the multi-modal approach to biometric system development in a perpetual quest for better recognition methods.

REFERENCES

Bartlett, M. S., Movellan, J. R., & Sejnowski, T. J. (2002). Face recognition by independent component analysis. *IEEE Transactions on Neural Networks*, *13*(6), 1450–1464. doi:10.1109/TNN.2002.804287 PMID:18244540.

Baudat, G., & Anouar, F. (2002). Generalized discriminant analysis using a kernel approach. *Neural Computation, 12*(10), 2385–2404. doi:10.1162/089976600300014980 PMID:11032039.

Bebis, G., Deaconu, T., & Georiopoulous, M. (1999). Fingerprint identification using Delaunay triangulation. In Proceedings of ICIIS99, (pp. 452-459). ICIIS.

Belhumeur, P., Hespanha, J., & Kriegman, D. (1997). Eigenfaces vs. fisherfaces: Recognition using class specific linear projection. *IEEE Transactions on Pattern Analysis and Machine Intelligence, 19*(7), 711–720. doi:10.1109/34.598228.

Cappelli, R., Maio, D., & Maltoni, D. (2001). Modelling plastic distortion in fingerprint images. *Lecture Notes in Computer Science, 2013*, 369–376. doi:10.1007/3-540-44732-6_38.

Cuntoor, N., Kale, A., & Chellappa, R. (2003). Combining multiple evidences for gait recognition. In *Proceedings of the 2003 International Conference on Multimedia and Expo*, (pp. 113-116). IEEE Computer Society.

Cutting, J. K., & Kozlowski, L. K. (1977). Recognizing friends by their walk: Gait perception without familiarity cues. *Bulletin of the Psychonomic Society, 9*(5), 353–356.

Daugman, J. (1993). High confidence visual recognition of persons by a test of statistical independence. *IEEE Transactions on Pattern Analysis and Machine Intelligence, 15*, 1148–1161. doi:10.1109/34.244676.

Daugman, J. (2004). How iris recognition works. *IEEE Transactions on Circuits and Systems for Video Technology, 14*(1), 21–30. doi:10.1109/TCSVT.2003.818350.

Delhi, I. I. T. (2012). *Ear database*. Retrieved from http://www4.comp.polyu.edu.hk/~csajaykr/IITD/Database_Ear.htm

Gabor, D. (2012). Theory of communication. *Journal of the Institute of Electrical Engineering, 93*, 429–457.

Gavrilova, M. (2004). Computational geometry and biometrics: On the path to convergence. In *Proceedings of the International Workshop on Biometric Technologies 2004*, (pp. 131-138). Calgary, Canada: IEEE.

Gavrilova, M., & Ahmadian, K. (2011). Dealing with biometric multi-dimensionality through novel chaotic neural network methodology. *International Journal of Information Technology and Management, 11*(1-2), 18–34.

Gavrilova, M., & Monwar, M. M. (2008). Fusing multiple matcher's outputs for secure human identification. *International Journal of Biometrics, 1*(3), 329–348. doi:10.1504/IJBM.2009.024277.

Hamming, R. W. (1950). Error detecting and error correcting codes. *The Bell System Technical Journal, 29*(2), 147–160.

Hough, P. V. C. (1962). *Method and means for recognizing complex patterns*. US Patent 3069654. Washington, DC: US Patent Office.

Jain, A. Flynn, P., & Ross, A. (2007). Handbook of biometrics. New York, NY: Springer.

Jain, A., Hong, L., & Bolle, R. (1997). On-line fingerprint verification. *IEEE Transactions on Pattern Analysis and Machine Intelligence, 4*, 302–313. doi:10.1109/34.587996.

Jiang, X., & Yau, W.-Y. (2000). Fingerprint minutiae matching based on the local and global structures. In *Proceedings of the 15th Internet Conference on Pattern Recognition (ICPR, 2000)*, (vol. 2, pp. 1042–1045). ICPR.

Johansson, G. (1973). Visual perception of biological motion and a model for its analysis. *Perception & Psychophysics, 14*(2), 201–211. doi:10.3758/BF03212378.

Kim, K. I., Jung, K., & Kim, H. J. (2002). Face recognition using kernel principal component analysis. *IEEE Signal Processing Letters*, *9*(2), 40–42. doi:10.1109/97.991133.

Kovacs-Vajna, Z., & Miklos, A. (2000). Fingerprint verification system based on triangular matching and dynamic time warping. *IEEE Transactions on Pattern Analysis and Machine Intelligence*, *22*(11), 1266–1276. doi:10.1109/34.888711.

Liu, J., & Zheng, N. (2007). Gait history image: A novel temporal template for gait recognition. In *Proceedings of the 2007 IEEE International Conf. on Multimedia and Expo*, (pp. 663-666). Beijing, China: IEEE.

Lu, J., Plataniotis, K. N., & Venetsanopoulos, A. N. (2003). Face recognition using LDA-based algorithms. *IEEE Transactions on Neural Networks*, *14*(1), 195–200. doi:10.1109/TNN.2002.806647 PMID:18238001.

Phillips, P. J. (1998). Support vector machines applied to face recognition. *Advances in Neural Information Processing Systems*, *11*, 113–123.

Preparata, F., & Shamos, M. (1985). *Computational geometry: An introduction*. Berlin, Germany: Springer. doi:10.1007/978-1-4612-1098-6.

Ratha, N. K., Karu, K., Chen, S., & Jain, A. (1996). A real-time matching system for large fingerprint databases. *IEEE Transactions on Pattern Analysis and Machine Intelligence*, *18*(8), 799–813. doi:10.1109/34.531800.

Rowley, H. A., Baluja, S., & Kanade, T. (1998). Neural network-based face detection. *IEEE Transactions on Pattern Analysis and Machine Intelligence*, *20*(1), 23–38. doi:10.1109/34.655647.

Soledek, J., Shmerko, V., Phillips, P., Kukharevl, G., Rogers, W., & Yanushkevich, S. (1997). Image analysis and pattern recognition in biometric technologies. In *Proceedings of International Conference on the Biometrics: Fraud Prevention, Enhanced Service*, (pp. 270-286). Las Vegas, NV: IEEE.

Turk, M., & Pentland, A. (1991). Eigenfaces for recognition. *Journal of Cognitive Neuroscience*, *3*(1), 71–86. doi:10.1162/jocn.1991.3.1.71.

Vacca, J. R. (2007). *Biometric technologies and verification systems*. Burlington, MA: Butterworth-Heinemann.

Wang, C., & Gavrilova, M. (2004). A multi-resolution approach to singular point detection in fingerprint images. In *Proceedings of the International Conference of Artificial Intelligence*, (vol. 1, pp. 506-511). IEEE.

Wang, C., & Gavrilova, M. (2005). A novel topology-based matching algorithm for fingerprint recognition in the presence of elastic distortions. In *Proceedings of the International Conference on Computational Science and its Applications*, (vol. 1, pp. 748-757). Springer.

Wang, C.-H. (2005). *A literature survey on human gait recognition techniques. Directed Studies EE8601*. Toronto, Canada: Ryerson University.

Wang, H., Gavrilova, M., Luo, Y., & Rokne, J. (2006). An efficient algorithm for fingerprint matching. In *Proceedings of International Conference on Pattern Recognition*, (pp. 1034-1037). IEEE.

Wang, J., She, M., Nahavandi, S., & Kouzani, A. (2010). A review of vision-based gait recognition methods for human identification. In *Proceedings of the 2010 Digital Image Computing: Techniques and Application*, (pp. 320-327). Piscataway, NJ: IEEE.

Wayman, J., Jain, A., Maltoni, D., & Maio, D. (2006). *Biometric systems: Technology, design and performance evaluation*. Berlin, Germany: Springer-Verlag.

Wecker, L., Samavati, F., & Gavrilova, M. (2005). Iris synthesis: A multi-resolution approach. In *Proceedings of 3rd International Conference on Computer Graphics and Interactive Techniques in Australasia and South East Asia*, (pp. 121-125). IEEE.

Xiao, Y., & Yan, H. (2002). Facial feature location with delaunay triangulation/voronoi diagram calculation. In Feng, D. D., Jin, J., Eades, P., & Yan, H. (Eds.), *Conferences in Research and Practice in Information Technology* (pp. 103–108). ACS.

Yanushkevich, S., Gavrilova, M., Wang, P., & Srihari, S. (2007). *Image pattern recognition: Synthesis and analysis in biometrics*. New York, NY: World Scientific Publishers.

Zhang, D. (2004). *Palmprint authentication*. Berlin, Germany: Springer.

Section 2
Current Practices in Information Fusion for Multimodal Biometrics

Chapter 4
Multimodal Biometric System and Information Fusion

ABSTRACT

Integrating different information originating from different sources, known as information fusion, is one of the main factors of designing a biometric system involving more than one biometric source. In this chapter, various information fusion techniques in the context of multimodal biometric systems are discussed. Usually, the information in a multimodal biometric system can be combined in senor level, feature extraction level, match score level, rank level, and decision level. There is also another emerging fusion method, which is becoming popular—the fuzzy fusion. Fuzzy fusion deals with the quality of the inputs or with the quality of any system components. This chapter discusses the associated challenges related to making the choice of appropriate fusion method for the application domain, to balance between fully automated versus user defined operational parameters of the system and to take the decision on governing rules and weight assignment for fuzzy fusion.

1. INTRODUCTION

The optimal biometric system is one having the properties of distinctiveness, universality, permanence, acceptability, collectability, and security. As we saw in the introductory chapters, no existing biometric security system simultaneously meets all of these requirements. Despite tremendous prog-ress in the field, over the last decades researchers noticed that while a single biometric trait might not always satisfy secure system requirements, the combination of traits from different biometrics will do the job. The key is in aggregation of data and intelligent decision making based on responses received from individual (unimodal) biometric systems.

DOI: 10.4018/978-1-4666-3646-0.ch004

Thus, Multimodal biometrics emerged as a new and highly promising approach to biometric knowledge representation, which strives to overcome problems of individual biometric matchers by consolidating the evidence presented by multiple biometric traits (Ross, Nandukamar, & Jain, 2006). As an example, a multimodal system may use both face recognition and signature to authenticate a person. Due to reliable and efficient security solutions in the security critical applications, multimodal biometric systems have evolved over last decade as a viable alternative to the traditional unimodal security systems.

2. ADVANTAGES OF MULTIMODAL BIOMETRIC SYSTEM

The advantages of multimodal biometric systems over unimodal systems are mainly due to utilization of more than one information source. Figure 1 shows a sample multimodal biometric system. The most prominent implications of this are increased and reliable recognition performance, fewer enrolment problems, and enhanced security (Ross & Jain, 2004).

2.1. Increased and Reliable Recognition Performance

A multimodal system allows for a greater level of assurance of a proper match in verification and identification modes w(Hong & Jain, 1998). As multimodal biometric systems use more than one biometric trait, each of those traits can offer additional evidence about the authenticity of any identity claim. For example, the gaits (the patterns of movements) of two persons of the same family (or coincidentally of two different persons) can be similar. In this scenario, a unimodal biometric system based only on gait pattern analysis may results in false recognition. If the same biometric system also includes fingerprint matching, the

system would results in increased recognition rate, as it is very unlikely that two different persons have same gait and fingerprint patterns.

Another example of increased and reliable recognition performance of multimodal biometric systems is ability to effectively handle the noisy or poor quality data. When the biometric information acquired from a single trait is not reliable due to noise, the availability of other trait allows the system to still perform in a secure manner. For example, in a face and voice-based multimodal biometric system, due to noise, if the individual's voice signals cannot be accurately measured, the facial characteristics may be used for authentication.

2.2. Fewer Enrolment Problems

Multimodal biometric systems address the problem of non-universality or the insufficient population coverage, where a portion of a population has a biometric characteristic that is missing or not suitable for recognition, and thus reduce the failure to enroll rate significantly (Frischholz & Dieckmann, 2000). Depending on the system design, many multimodal biometric systems can perform matching even in the absence of one of the biometric samples. For example, in a fingerprint and face based multimodal system, a person (who is a carpenter) cannot enroll his fingerprint information to the system due to the scars in his fingerprint. In this case, the multimodal system can still perform authentication using the facial characteristics of that person. Moreover, if certain features can be extracted from fingerprint (but not all due to damage to the finger), then these features still can be sued to increase accuracy rate or confidence level of the final decision.

2.3. Enhanced Security

Multimodal biometric systems make it more difficult for an impostor to spoof biometric traits of

Figure 1. A sample multibiometric system architecture

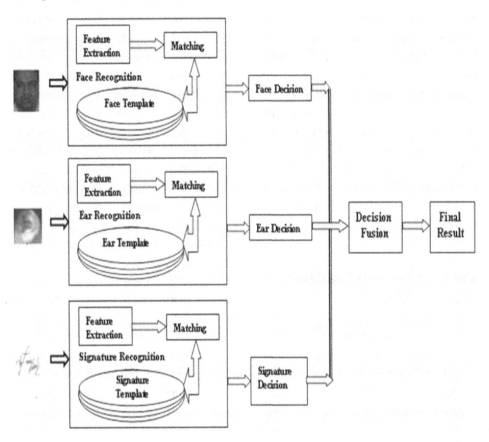

a legitimately enrolled individual. A spoof attack is where a person pretends to be another person by using stolen ID or falsified information. For example, researchers have demonstrated how to create fake fingerprints, which had some success in bypassing commercial fingerprint recognition system security (Matsumoto, Matsumoto, Yamada, & Hoshino, 2002). The advantage of multimodal systems is that the impostor would need to spoof more than one biometric trait simultaneously, which would be significantly more challenging.

Multimodal biometric systems can also serve as a fault tolerant system. For example, multimodal systems can still perform their functions and result in relatively reliable outcomes even when certain biometric modules stop operating (due to sensor malfunction, software issues, unavailability of a sample data or quality being extremely low). The more high quality data is received, however, the better overall multibiometric system accuracy rates would normally become.

3. DEVELOPMENTAL ISSUES OF MULTIBIOMETRIC SYSTEMS

Development of a multibiometric system for security purposes is not a trivial task. As with any unimodal system, the data acquisition procedure, sources of information, level of expected accuracy, system robustness, user training, data privacy, and dependency on proper functioning of hardware and proper operational procedures impact directly the performance of security system. While using more than one data source alleviates some issues (such as noisy data, missing samples, errors in acquisition, spoofing etc.), this advantage does not come free. The choice of biometric information that needs to be integrated or fused must be made, information fusion methodology should be selected, cost vs. benefit analysis needs to be performed, processing sequences developed, and system operators trained.

We now focus on these issues specifically in the context of multibiometric systems.

3.1. Ease of Data Acquisition Procedure

One of the key design issues of a multibiometric based security system is a convenient interface with the system to ensure the efficient acquisition of biometric information. As stated in Oviatt (2003): "An appropriately designed interface can ensure that multiple pieces of evidence pertaining to an individual's identity are reliably acquired whilst causing minimum inconvenience to the user". For example, in a face, ear and fingerprint based multimodal biometric system, if a user needs to present his/her three biometric identifiers separately, that would be very inconvenient. Instead, if the three biometric identifiers can be acquired simultaneously (or in one seating), that might be more convenient for user. Unfortunately, up to date, there have been too few studies looking into this aspect of human-computer interaction with the biometric system.

3.2. Source of Information

Multibiometric systems are based on more than one biometric information source. Multiple biometric information can be derived from multiple identifiers, from single identifiers but with multiple samples or instances, or from a combination of both (Ross, Nandukamar, & Jain, 2006). The sources of biometric information in such systems depend on a variety of issues including the ap-plication necessity and scenarios, the availability of the biometric information, the cost associated with the biometric information acquisition process, the choice of pattern matching and information fusion algorithms.

3.3. Choice of Biometric Information

Biometric information integration (or fusion) can occur in various levels, from the initial stage after acquiring the raw data to the final stage after obtaining the final match/non-match decision. Next, the extracted features, the matching scores or the final ranked list—all of these can be integrated in a multibiometric system. Which information needs to be fused is one of the crucial decisions of a multibiometric system design (Ross, Nandukamar, & Jain, 2006). The integration usually depends on the application scenario and on the availability of information. For example, in some multibiometric systems (specially, in commercial biometric security systems) only the final decision is available. In this case, only the decision fusion is possible for that multibiometric system.

3.4. Information Fusion Methodology

For all types of information fusion in multibiometric systems, there are several alternative algorithms that can be used (Ross, Nandukamar, & Jain, 2006). For example, to get the consensus ranked lists, the initial ranked lists (obtained after matching input and templates) can be integrated by the highest rank method, Borda count method, logistic regression method, Bayesian method, fuzzy method, or Markov chain method. What approach needs to be taken depends on the designer of the system, previous performances of the methodologies and the robustness required for the system.

3.5. Cost vs. Benefits

One of the drawbacks of multibiometric system development is the higher cost, as compared to a single biometric based security system. So, before making commitment to multimodal biometric approach, analyzing the potential benefits that can be obtained through the development of multibiometric system is necessary. The cost depends on the number of sensors deployed, the time it takes to acquire biometric data, user or system operator experience, and system maintenance (Ross, Nandukamar, & Jain, 2006).

3.6. Processing Sequences

Another important issue in multibiometric system design is how the acquisition or data processing will occur for the system (Ross, Nandukamar, & Jain, 2006). Should the data be acquired or processed simultaneously, or should the data be acquired or processed in parallel must be decided in advance. Usually, there are two possible ways to choose the data acquisition sequences. In serial data acquisition process, the multibiometric data is acquired sequentially and within a short time interval (Ross, Nandukamar, & Jain, 2006). In the parallel data acquisition process, all the multibiometric data is gathered in parallel which makes the system faster than sequential method (Ross, Nandukamar, & Jain, 2006).

In the data processing stage, parallel or cascade mode can be used in any multibiometric system. In the cascade mode, the processing of biometric data occurs sequentially, while in parallel biometric data processing, all biometric data is processed simultaneously and used in authentication process (Ross, Nandukamar, & Jain, 2006).

Figure 2 illustrates cascade processing sequence and Figure 3 illustrates parallel processing sequences for multibiometric security system.

4. INFORMATION SOURCES FOR MULTIBIOMETRIC SYSTEMS

For many applications, there are additional sources of non-biometric information that can be used for person authentication, while in others the use of a single biometric is not sufficiently secure or does not provide adequate coverage of the user population (Bolle, Connell, Pankanti, Ratha, & Senior, 2004). This can be indicated by such parameter as Failure to Enroll rate. Thus, multibiometric system emerged as a way to provide more secure and reliable person authentication system under those conditions (Bolle, Connell, Pankanti, Ratha, & Senior, 2004).

It must be pointed out that in literature there is a slight difference between two terms. The term *multimodal* biometric system refers specifically to those biometric systems where more than one biometric modalities are used (Ross, Nandukamar, & Jain, 2006). The term *multibiometric* is more generic and includes multimodal systems and some other configurations using only one biometric modality with different samples instances or algorithms (Soltane, Doghmane, & Guersi, 2010).

Multiple Sensors - One Biometric Trait: In these systems, different sensors are used for capturing different representations of the same biometric modality to extract diverse information (Ross, Nandukamar, & Jain, 2006). For example, a biometric system may use 2D, 3D, or thermal face image for authentication. As these systems consider only one biometric trait, so if the specific biometric trait is missing or not suitable, the performance benefits of multiple acquisitions will be minimal.

Multiple Instances - One Biometric Trait: In these systems, multiple instances of the same biometric trait are used for authentication (Ross, Nandukamar, & Jain, 2006). For example, the image of left and right eye of a subject may be used for retina recognition system. These systems are cost efficient, as the same sensors or the same feature extraction and matching algorithm can be used.

Multiple Algorithms – One Biometric Trait: These systems use one biometric trait but use different matching algorithms (Ross, Nandukamar, & Jain, 2006). For example, a system may use eigenface and Voronoi diagram as matching algorithms for the same set of face images and

Figure 2. Multimodal biometric data processing sequence: cascade mode (adopted from Ross, Nandu-kamar, & Jain, 2006)

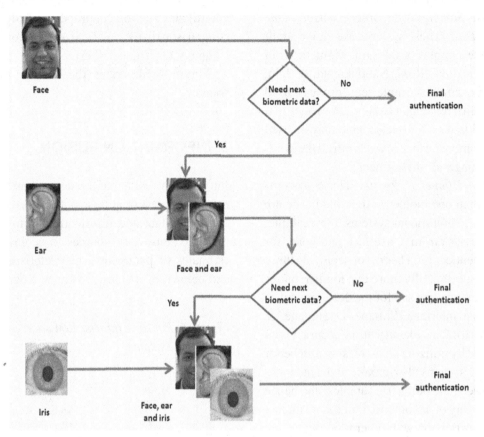

Figure 3. Multimodal biometric data processing sequence: parallel mode (adopted from Ross, Nandu-kamar, & Jain, 2006)

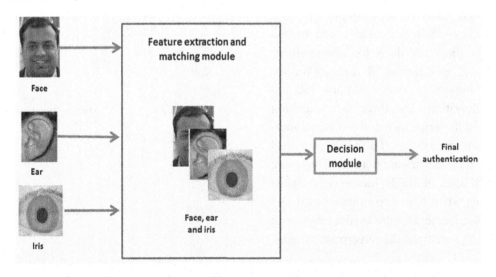

later combine the results. These systems also suffer with the poor quality of input.

Multiple Samples with Single Sensor– One Biometric Trait: These systems use single senor but multiple samples of the same biometric trait for authentication (Ross, Nandukamar, & Jain, 2006). For example, a single sensor may be used to capture different facial expression images of a subject and latter a mosaic scheme may be used to build a composite face image from all the available face images of that subject.

Multiple Biometric Traits: These systems use more than one biometric trait and hence are referred to as multimodal systems. For example, a biometric system may use face and voice for person authentication. The cost of deploying these systems is substantially more due to the requirements of new sensors and for the development of the new user interface (Soltane, Doghmane, & Guersi, 2010). The identification accuracy can be improved by utilizing an increasing number of traits. These systems also maximize the independence between the biometric samples, and hence the poor quality of a biometric trait has no impact on the authentication with other trait.

Multiple Tokens: This is the typical authentication system consists of one or more biometric identifiers as well as a possession or knowledge token (Bolle, Connell, Pankanti, Ratha, & Senior, 2004). The possession and knowledge token can be ID card and password, for example.

Hybrid Systems: These systems use more than one scenarios discussed above for robust authentication (Ross, Nandukamar, & Jain, 2006). For example, a biometric system may use two iris matching algorithms and three face matching algorithms in the same face and iris based multimodal biometric system. The ideas of hybrid algorithms in biometrics are not new. They were successfully used in single biometric recognition systems, when both appearance-based and topology-based methods were used for enhanced recognition. As example, the fingerprint recogni-

tion system based on Voronoi diagram relies both on geometric properties (such as triangle edge length) and topological properties (ridge pattern comparison) in making final recognition decision (Wang & Gavrilova, 2005).

Figure 4 illustrates the various biometric sources.

5. INFORMATION FUSION

Information fusion can be defined as "an information process that associates, correlates and combines data and information from single or multiple sensors or sources to achieve refined estimates of parameters, characteristics, events and behaviors" (Linas, Bowman, Rogova, Sein-

Figure 4. Possible information sources of multibiometric systems

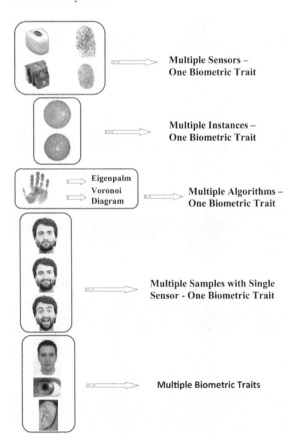

berg, Waltz, & White, 2004). A good information fusion method allows the impact of less reliable sources be lowered compared to reliable ones. A number of disparate research areas including robotics, image processing, pattern recognition, information retrieval etc. utilize and describe information fusion in their context. Thus, information fusion established itself as an independent research area over the last decade for its impact on a vast number of disparate research areas. For example, the concept of *data and feature fusion* initially occurred in multi-sensor processing. In fact, information fusion was for a long time used in engineering and signal processing fields, as well as in decision-making and expert systems. By now, several other research fields found its application

useful. Besides the more classical data fusion approaches in robotics, image processing and pattern recognition, the information retrieval community has been known to combine multiple information sources (Wu & McClean, 2006). Figure 5 shows the basic building block of an information fusion system which fuses source information at the early stage of the system.

The origins of *classifier and decision fusion* can be traced back to neural network literature, where the method combining neural network outputs was published in 1965 (Tumer & Gosh, 1999). Later, its application was expanded into other fields like economics for forecast combining, machine learning for evidence combination and also information retrieval for page rank aggrega-

Figure 5. Block diagram of a sensor information fusion system

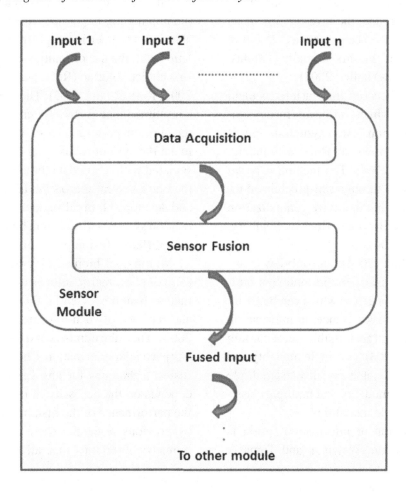

tion (Wu & McClean, 2006). Since early application areas of data, classifier and decision fusion, researchers were wondering about which level of information fusion is to be preferred and, more generally, how to design an optimal information fusion strategy for multimedia processing systems (Kludas, Bruno, & Marchand-Maillet, 2008). Examples using classifier fusion are numerous, including multimedia retrieval (Wu, Chang, Chang, & Smith, 2004), multi-modal object recognition (Wu, Cohen, & Oviatt, 2002), multibiometrics (Poh & Bengio, 2005), and video retrieval (Yan & Hauptmann, 2003). The applications of decision fusion can be found in multimedia (Benitez & Chang, 2002), text and image categorization (Chechik & Tishby, 2003), multi-modal image retrieval (Westerveld & de Vries, 2004), and Web-based document retrieval (Zhao & Grosky, 2002). Determination of fusion performance improvement and the investigation of its influence factors have been identified as future research directions in this domain (Kludas, Bruno, & Marchand-Maillet, 2008).

The connection between drawing inference and information fusion, which is utilizable for automation of the information fusion system design, is presented in Linas, Bowman, Rogova, Seinberg, Waltz, and White (2004). The method is based on pattern discovery through self-proclaimed innovative reasoning and inductive generalization approaches (Linas, Bowman, Rogova, Seinberg, Waltz, & White, 2004).

In Ross and Jain (2004), an overview of information fusion scenario in the context of multimodal biometrics is given, which can be easily adapted to general tasks. Hence, in information fusion the settings that are possible are, according to Ross and Jain (2004): (1) single modality and multiple sensors, (2) single modality and multiple features, (3) single modality and multiple classifiers, and (4) multiple modalities.

The interpretation of information fusion is documented in Kokar, Weyman, and Tomasik (2004): "By combining low level features, it is possible to achieve a more abstract or a more precise representation of the world." The Durrant-Whyte classification of information fusion strategies is presented in Linas, Bowman, Rogova, Seinberg, Waltz, and White (2004). Optimization of a fusion system is considered in Fassinut-Mombot and Choquel (2004). Also, in the presence of higher reliable biometric unimodal system, it is possible to adjust the confidence and thus assign a higher importance to a more reliable source (Aarabi & Dasarathy, 2004).

6. BIOMETRIC INFORMATION FUSION

Due to some problems associated with the unimodal biometric data, such as small variation over the population, large intra-variability over time, absence of biometric sample in portion of a population etc., the use of multimodal biometrics is a first choice solution (Ross, Nandukamar, & Jain, 2006; Ross & Jain, 2003). The main objective of a multimodal biometric system is to improve the recognition performance of the system and to make the system robust over the limitations associated with unimodal biometric systems. Over the years, several approaches have been proposed and developed for multimodal biometric authentication system with different biometric traits and with different fusion mechanisms.

Multimodal biometric systems use multiple sources of biometric information, whereas information fusion is essential for analysis, indexing and retrieval of such information (Ross & Jain, 2003). There are numbers of fusion techniques for any particular information. Choosing appropriate fusion techniques for any specific information depends on the necessity of the application and the performance of the fusion techniques proven by previous research. There is a consensus in biometric literature that all various levels of multimodal biometric information fall into two broad categories: before matching and after matching fusion (Sanderson & Paliwal, 2001). Fusion before matching category contains *sensor level fusion* and *feature level fusion*, while fusion after matching contains *match score level fusion*, *rank level fusion* and *decision level fusion* (Sanderson & Paliwal, 2001). A novel fusion mechanism has been established recently in BTLab is based on *fuzzy logic* fusion, and hence named a *fuzzy biometric fusion* (Monwar, Gavrilova, & Wang, 2011). Fuzzy biometric fusion can be employed either in the initial stage, i.e. before matching occurred or in the latter stage, i. e. after matching occurred.

Figure 6 shows the multimodal biometric fusion classification. Figure 7 shows the possible fusion before matching and fusion after matching levels.

7. FUSION BEFORE MATCHING AND FUSION AFTER MATCHING

Fusion in this category integrates evidences before matching or comparison of data samples against the user sample occurs. According to Kokar et al., "By combining low level features it is possible to achieve a more abstract or a more precise representation of the world" (Kokar, Weyman, & Tomasik, 2004). Thus, biometric sources at the earlier stage contain much more information than after processing) (Ross & Jain, 2003).

However, the extra costs of storing raw data, and additional complexity in developing matching methods do not make this approach quite practical.

Fusion after-matching methods consolidate information obtained after individual biometric matching or comparison is done (Ross & Jain, 2003). Most multimodal biometric systems have been using these fusion methods as the information needed for fusion is easily available compared to fusion before matching methods. The matching scores, the ranking list (sorted order) based on matching scores or the individual biometric

decision (Yes/No) can be used for fusion in this category.

7.1. Sensor Level Fusion

In this level of fusion, the raw data acquired from multiple sensors can be processed and integrated to generate new aggregated data from which features can be extracted. Sensor level fusion is used where the same biometric identifiers are collected using multiple sensors or when multiple samples of the same biometric trait are captured using a single sensor (Ross, Nandukamar, & Jain, 2006; Jain, 2005). For example, in the case of face biometrics, color, geometry, depth, and texture information may be combined to generate a 3D textured image of a face (Hsu, 2002).

Following a similar idea, a face mosaicing technique was proposed in Liu and Chen (2003). This is a method for combining two or more images of the same face. The authors used a 3D ellipsoidal model to approximate the human head images and validated their method on CMU facial database (Sim, Baker, & Bsat, 2003).

Another contribution in this area is the research reported in (Raghavendra, Rao, & Kumar, 2010). They proposed an approach to combine information obtained from face and palmprint image using Particle Swarm Optimization (PSO). The Kernel Direct Discriminant Analysis (KDDA) was used for the final classification decision. Using FRGC face database (Phillips, et al., 2005) and POLYU palmprint database (Jing, Yao, Yang, Li, & Zhang, 2007), authors tested their recognition performance with match score level fusion and with genetic algorithm applied on the same set of databases.

7.2. Feature Level Fusion

Feature level fusion combines more than one feature sets extracted from multiple data sources. The geometric features of the face, for example, may be combined with eigenvectors in order to

Figure 6. Biometric fusion classification

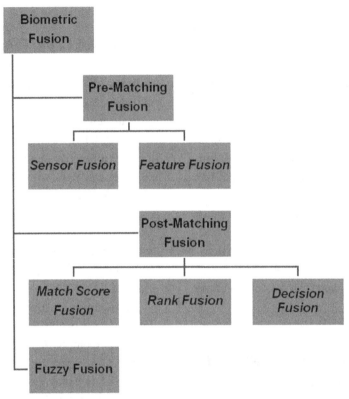

Figure 7. Possible fusion before matching and fusion after matching levels

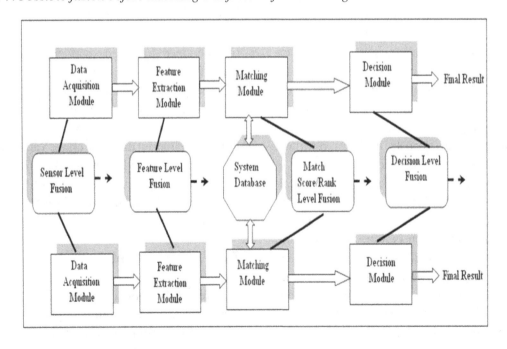

construct a new high-dimension feature vector (Ross & Govindarajan, 2005).

Feature level fusion can be achieved by template perfection or template enhancement of the extracted feature. Template perfection can be done based on the evidence presented by the current feature set in order to reflect permanent changes (if any) in a person's biometric identifier (Ross & Jain, 2003). Template enhancement can be done by simply combining two feature sets of the same biometric identifier of a person into one (Ross & Jain, 2003).

This fusion method is expected to produce comparatively better results than other fusion methods as more raw information is available for fusion which may be unavailable in after- matching fusion methods. But there are some difficulties if the feature sets originate from various biometric traits (Ross, Nandukamar, & Jain, 2006). The feature sets from different modalities may be obtained by different algorithms, thus training the system to perform accurate recognition may be more difficult. Also, this type of fusion creates 'curse of dimensionality' problem, which is known as problem associated with high-dimensional features space. The problem in turn needs to be resolved through some dimensionality reduction techniques (such as space transformations, clustering, etc.). In addition, a feature normalization technique may be necessary if the feature sets exhibit significant differences in their range as well as distribution (Jain, 2005). Finally, in most commercial biometric systems, feature sets are kept confidential and their processing is not allowed. On the other hand, if the feature sets originate from single biometric identifier, template update or template improvement algorithms can be used (Moon, Yeung, Chan, & Chan, 2004).

One example of this fusion level is a multimodal system for face and palmprint (Feng, Dong, Hu, & Zhang, 2004). The system used and compared two common methods: Principal Component Analysis (PCA) and Independent Component Analysis (ICA) As noted by authors, ICA performed bet-

ter than PCA in both unimodal and multimodal validation framework.

In another attempt to develop a feature fusion multimodal biometric system, Rattani et al. (Rattani, Kisku, Bicego, & Tistarelli, 2010) proposed a multimodal biometric system combining face and fingerprint information at the feature level. The authors implemented several feature reduction techniques. They conducted experiments on BANCA face database (Bailly-Baillire, et al., 2003) and a local fingerprint database to evaluate the recognition accuracy with match score level fusion for the same data.

7.3. Match Score Level Fusion

Match score level fusion method consolidates matching scores generated from different classifiers and can be applied to most of the multibiometric scenarios (He, et al., 2010). For example, this fusion method can combine matching scores obtained from two different algorithms for two instances of fingerprints. This fusion method can also be used to consolidate matching scores obtained from a face matcher and an iris matcher.

For obtaining a single matching score, match score fusion applies arithmetic operations, such as addition, subtraction, maximum, minimum, or median, to different matching scores (Ross & Jain, 2003). As an example, the match scores generated by three different matchers for the face, fingerprint and hand may be combined via the simple sum rule in order to obtain a new match score which is then used to make the final decision (Ross & Jain, 2003). As different matching scores from different algorithms may not share the same underlying properties or the score range, the score normalization is necessary in match score level fusion methods. Min-max, decimal scaling, z-score, median, median absolute deviation, double sigmoid are some examples of score normalization techniques. Normalization process is costly in terms of time and choosing inappropriate normalization can lead to very poor recognition

Figure 8. Match score level fusion for a three classifiers based multimodal biometric system

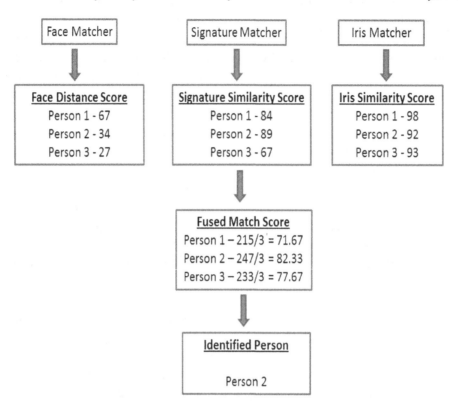

accuracy. Figure 8 illustrates match score level fusion for a multimodal biometric system.

In 1998, a bimodal approach was proposed for a Principal Component Analysis (PCA) based face and minutiae-based fingerprint identification system with a fusion method at the match score level (integrating the matching scores of different classifiers and making a decision based on the consensus matching scores) (Hong & Jain, 1998). The paper used MSU fingerprint database (Jain, Hong, & Bolle, 1997) for validation. The results were highly encouraging and demonstrated very good performance in terms of false accept rate and false reject rate.

In 2005, a multimodal approach for face, fingerprint and hand geometry, with fusion at the score level was proposed (Jain, Nandakumar, & Ross, 2005). The matching approaches for these modalities are minutiae-based matcher for fingerprint, which has similarity scores as output,

PCA-based algorithm for face recognition, which has Euclidean distance as output, and a 14-dimensional features vector for hand-geometry, which also has Euclidean distance as output. Seven score normalization techniques (simple distance-t-similarity transformation with no change in scale, min-max normalization, z-score normalization, median-MAD normalization, double sigmoid normalization, Tanh normalization and Parzen normalization) and three fusion techniques on the normalized scores (simple-sum-rule, max-rule and min-rule) were tested in this study. Except for the median-MAD normalization technique, all fusion approaches outperform the unimodal approaches with a significant increase in recognition accuracy.

7.4. Rank Level Fusion

Rank level fusion consolidates multiple ranking lists obtained from several biometric matchers to form a final ranking list, which would aid in establishing the final decision (Ross, Nandukamar, & Jain, 2006). Some industrial biometric devices often output only the ranked lists with user identities. In those cases, information on matching scores or features may not be available. Furthermore, in some biometric systems, the matching scores from the matchers are not suitable for subsequent fusion (Kumar & Shekhar, 2010). Therefore, the decision on confidence in user identification from multiple sources is made by rank level combination method. Techniques such as Borda count (Borda, 1781) may be used to make the final decision (Black, 1963). Methods for rank level fusion are discussed in details in the chapters seven and eight.

7.5. Decision Level Fusion

Decision level fusion method consolidates the final decision of single biometric matchers to form a final aggregated decision. When each matcher outputs its own class label (i.e., accept or reject in a verification system, or the identity of a user in an identification system), a single class label can be obtained by employing techniques, such as, "AND"/"OR", majority voting, weighted majority voting, decision table, Bayesian decision and Dempster-Shafer theory of evidence (Jain, 2005). The combination schemes are considered to be less complex than for other modes of fusion (Ross, Nandukamar, & Jain, 2006). This fusion method is suitable for those commercial biometric systems where only the final decision is available (Ross, Nandukamar, & Jain, 2006).

In 2000, a commercial multimodal approach, BioID, was developed for a model-based face classifier, a vector quantization (VQ)-based voice classifier and an optical-flow-based lip movement classifier for verifying persons (Frischholz & Dieckmann, 2000). Lip motion and face images were extracted from a video sequence and the voice from an audio signal. Weighted sum rule and majority voting approaches of decision level fusion method were used for fusion. Their experi-

ments demonstrated that the system can reduce the FAR and guarantee high reliability.

In another work, a multibiometric approach which combines palmprint, fingerprint and finger geometry collected by a digital camera at decision fusion level was presented (Yu, Xu, Zhou, & Li, 2009). Three decision fusion rules, including "AND" rule, "OR" rule and majority voting, are employed to perform the fusion. Among the three decision fusion methods, majority voting performed better than the other two decision methods.

7.6. Fuzzy Fusion

Fuzzy logic based fusion is another impressive information fusion approach which has been successfully applied in many different applications for the past years, such as automatic target recognition, medical image fusion and segmentation, gas turbine power plants fusion, weather forecasting, aerial image retrieval and classification, vehicle detection and classification and path planning (Monwar, Gavrilova, & Wang, 2011). With the fuzzy logic based fusion, one can obtain the level of confidence for the final recognition outcome which can be very important in some security critical biometric applications.

Here, it is assumed that the fuzzy fusion method can be employed in both before matching or after matching stages. When this fusion method is applied in before matching stage, usually it is to reduce the size of the dataset for comparison or matching. This fusion can also be employed in after matching stage to increase the recognition performance and to obtain the level of confidence of the final outcomes.

The method is based on fuzzy logic (Zadeh, 1965), which is the classic and most widely applied technology in computational intelligence (Wang, 2009). The fuzzy logic approach enables imprecise information to be processed in a way that resembles human thinking, e.g. big versus small, high versus low. It allows intermediate values to be defined between a normalized scope of {0, 1} by a partial membership for a fuzzy set. According to (Mitra & Pal, 2005), the significance of fuzzy logic in pattern recognition is present in the following areas:

- Representing linguistic input features for processing.
- Providing an estimation of missing information in terms of membership values.
- Representing ambiguous patterns and generating inferences in linguistic form.

The applications of fuzzy fusion are diverse. In Solaiman, Pierce, and Ulaby (1999), authors proposed a fuzzy-based multisensor data fusion classifier which was applied to land cover classification. The application domain here is the multisensor and contextual information fusion in a homogeneous framework. Due to the use of fuzzy concepts, the proposed classifier was ideally suited for integrating multisensor and a priori information.

In another study, authors developed a new vehicle classification algorithm using fuzzy logic (Kim, Kim, Lee, & Cho, 2001). In their algorithm, the vehicle weight and speed were used as the input to the fuzzy logic module. The output of the fuzzy logic module was a weighting factor to modify the vehicle length. The modified length was in turn the input to the vehicle classification module. The obtained results demonstrated that the proposed classification algorithm using the fuzzy logic reduces the errors in vehicle classification.

Another key contribution to the fuzzy fusion domain of literature is the work of Wang et al. (Wang, Dang, Li, & Li, 2007), where authors used fuzzy fusion for multimodal medical image application. To overcome the problem of blurriness of the most medical images, the authors proposed a new method of medical image fusion using fuzzy radial basis function neural networks (Fuzzy-RBFNN), which is functionally equivalent to T-S fuzzy model (Hunt, 1996). Genetic algorithm was used to train the network. Experiment

results, conducted on 20 groups of CT (Computed Tomography) and MRI (Magnetic Resonance Imaging) images of the head, showed that the proposed approach outperformed other methods in both visual effect and objective evaluation criteria, including experiments conducted on blurry images.

A 2010 paper (Deng, Su, Wang, & Li, 2010), proposed a data fusion method based on fuzzy set theory and Dempster-Shafer evidence theory (Shafer, 1976) for automatic target recognition. The authors represented both the individual attribute of a target in the model database and the sensor observation as fuzzy membership function and constructed a likelihood function to deal with fuzzy data collected with each sensor. Sensor data from different sources was used based on the Dempster combination rule (Dempster, 1976).

In another research on applying fuzzy fusion in the medical imaging research domain, Chaabane and Abdelouahab in 2011 (Chaabane & Abdelouahab, 2011) proposed a system of fuzzy information fusion framework for the automatic segmentation of human brain tissues. Their method consisted of the computation of fuzzy tissue maps in both images by using Fuzzy C-means algorithm, creation of fuzzy maps by a combination operator and a segmented image computation in a decision map.

Most recently, the application of fuzzy fusion method in biometric domain was presented in (Monwar, Gavrilova, & Wang, 2011). In that research, the fuzzy logic was employed to make decision in multi-modal biometric system based on face, ear, and signature. The rank level fusion method was extended with fuzzy logic decision making module, which operated on unimodal biometric data and was corresponding to the confidence level over results of individual matchers. On a second level, the results of rank level decision making module were fused and reported on a continuous scale (instead of binary Yes or No) as aggregated confidence in resulting outcome. Obtained results affirmed the method feasibility and high performance and emphasized the high potential of the data fusion in the biometric multimodal system domain. There are discussed in more details in Chapter 7.

Table 1 summarizes the review and shows different types of multibiometric systems for person authentication with different biometric identifiers and with different fusion approaches.

8. SUMMARY

In this chapter, information fusion techniques applied in multimodal biometrics area are discussed. Usually, the information originated from different sources in a multimodal biometric system can be combined in senor level, feature extraction level, match score level, rank level, and decision level. Among all of the fusion methods, senor fusion and feature extraction level fusion considered as the stage for combining raw data or the actual biometric data. Match score, rank and decision level fusion methods combine processed data or data obtained through some experimentations. There is also another novel fusion method which is becoming highly popular: the fuzzy fusion.

There are a number of challenges in this area, requiring further investigation. The first one is rooted in the choice of a fusion method, most appropriate for the application domain. The decision is often made ad-hoc, or based on non-essential constraints such as availability of the fusion module, low cost, etc, instead of being made based on actual fit of the application area and the method.

The second challenge lies in the balance between fully automated vs user defined operational parameters of the system. While complete automation might be a desired feature for some high0-demand large-scale applications, in practice this is not always possible or desirable. The best way to develop a biometric security system is to design it as a *decision-support system,* which can provide information to the system operator empowering him to make an intelligent and correct decision.

Table 1. Existing multimodal biometric systems

Year	Modalities Used for Fusion	Authors	Fusion	Fusion Approach
1998	Face and fingerprint	Hong and Jain (1998)	Match score	Product rule
2000	Face, voice and leap movement	Frischholz and Dieckmann (2000)	Decision	Weighted sum rule, Majority voting
2003	Face, fingerprint and hand geometry	Ross and Jain (2003)	Match Score	Sum-rule, decision tree and linear discriminant function
2003	2D and 3D faces	Liu and Chen (2003)	Sensor	Face mosaic
2004	Face and palmprint	Feng et al. (Feng, Dong, Hu, & Zhang, 2004)	Feature	Feature concatenation
2005	Face, fingerprint and hand geometry	Jain et al. (Jain, Nandakumar, & Ross, 2005)	Match score	Simple-sum-rule, max-rule and min-rule
2009	Fingerprint, face and hand geometry	Nandakumar et al. (Nandakumar, Chen, Dass, & Jain, 2009)	Match score	Likelihood ratio
2009	Hand biometrics (palmprint, fingerprint, finger geometry)	Yu et al. (Yu, Xu, Zhou, & Li, 2009)	Decision	AND rule, OR rule, majority voting
2010	Fingerprint and face	Rattani et al. (Rattani, Kisku, Bicego, & Tistarelli, 2010)	Feature	Delaunay triangulation
2010	Face and palmprint	Raghavendra et al. (Raghavendra, Rao, & Kumar, 2010)	Sensor	Particle swarm optimization
2010	Two palmprint images	Kumar and Shekhar (2010)	Rank	Borda count, weighted Borda count, maximum rank, nonlinear weighted rank

The final challenge is related to the fuzzy fusion method as one of the novel biometric research areas. The decision on governing rules and weight assignment in this method can significantly impact the outcomes of recognition process. Thus, the empirical validation of the system becomes crucial if this methodology is to be used. More discussion on that topic would be found in subsequent chapters.

REFERENCES

Aarabi, P., & Dasarathy, B. V. (2004). Robust speech processing using multi-sensor multi-source information fusion: An overview of the state of the art. *Information Fusion*, *5*, 77–80. doi:10.1016/j.inffus.2004.02.001.

Abaza, A., & Ross, A. (2009). Quality based rank-level fusion in multibiometric systems. In *Proceedings of 3rd IEEE International Conference on Biometrics: Theory, Applications and Systems*. Washington, DC: IEEE.

Ailon, N. (2007). Aggregation of partial rankings, p-ratings and top-m lists. *Algorithmica*, *57*(2), 284–300. doi:10.1007/s00453-008-9211-1.

Ailon, N., Charikar, M., & Newman, A. (2005). Aggregating inconsistent information: Ranking and clustering. In *Proceedings of 37th Annual ACM Symposium on Theory of Computing (STOC)*, (pp. 684-693). Baltimore, MD: ACM.

Bailly-Baillire, E., Bengio, S., Bimbot, F., Hamouz, M., Kittler, J., & Marithoz, J. Thiran, J. P. (2003). The BANCA database and evaluation protocol. In *Proceedings of International Conference on Audio- and Video-Based Biometric Person Authentication*, (pp. 625-638). Guildford, UK: IEEE.

Benitez, A. B., & Chang, S. F. (2002). Multimedia knowledge integration, summarization and evaluation. In *Proceedings of Workshop on Multimedia Data Mining*, (vol. 2326). Springer.

Black, D. (1963). *The theory of committees and elections* (2nd ed.). Cambridge, UK: Cambridge University Press.

Bolle, R. M., Connell, J. H., Pankanti, S., Ratha, N. K., & Senior, A. W. (2004). *Guide to biometrics*. New York, NY: Springer-Verlag.

Borda, J. C. (1781). *M'emoire sur les 'elections au scrutin*. Paris, France: Histoire de l'Acad'emie Royale des Sciences.

Chaabane, L., & Abdelouahab, M. (2011). Improvement of brain tissue segmentation using information fusion approach. *International Journal of Advanced Computer Science and Applications*, *2*(6), 84–90.

Chechik, G., & Tishby, N. (2003). Extracting relevant structures with side information. *Advances in Neural Information Processing Systems, 15*.

Dempster, A. P. (1976). Upper and lower probabilities induced by a multivalued mapping. *Annals of Mathematical Statistics*, *38*(2), 325–339. doi:10.1214/aoms/1177698950.

Deng, Y., Su, X., Wang, D., & Li, Q. (2010). Target recognition based on fuzzy Dempster data fusion method. *Defence Science Journal*, *60*(5), 525–530.

Fagin, R. (1999). Combining fuzzy information from multiple systems. *Journal of Computer and System Sciences*, *58*(1), 83–99. doi:10.1006/jcss.1998.1600.

Farah, M., & Vanderpooten, D. (2008). An outranking approach for information retrieval. *Information Retrieval*, *11*(4), 315–334. doi:10.1007/s10791-008-9046-z.

Fassinut-Mombot, B., & Choquel, J. B. (2004). A new probabilistic and entropy fusion approach for management of information sources. *Information Fusion, 5*, 35–47. doi:10.1016/j.inffus.2003.06.001.

Feng, G., Dong, K., Hu, D., & Zhang, D. (2004). When faces are combined with palmprint: A novel biometric fusion strategy. In *Proceedings of First International Conference on Biometric Authentication*, (pp. 701-707). Hong Kong, China: IEEE.

Frischholz, R., & Dieckmann, U. (2000). BioID: A multimodal biometric identification system. *IEEE Computer, 33*(2), 64–68. doi:10.1109/2.820041.

He, M. et al. (2010). Performance evaluation of score level fusion in multimodal biometric systems. *Pattern Recognition, 43*(5), 1789–1800. doi:10.1016/j.patcog.2009.11.018.

Hong, L., & Jain, A. K. (1998). Integrating faces and fingerprints for personal identification. *IEEE Transactions on Pattern Analysis and Machine Intelligence, 20*(12), 1295–1307. doi:10.1109/34.735803.

Hsu, R.-L. (2002). *Face detection and modeling for recognition*. (PhD Thesis). Michigan State University. East Lancing, MI.

Hunt, K. J. (1996). Extending the functional equivalence of radial basis function networks and fuzzy inference system. *IEEE Transactions on Neural Networks, 13*, 776–778. doi:10.1109/72.501735.

Jain, A. K. (2005). Biometric recognition: How do I know who you are? *Lecture Notes in Computer Science, 3617*, 19–26. doi:10.1007/11553595_3.

Jain, A. K., Hong, L., & Bolle, R. (1997). On-line fingerprint verification. *IEEE Transactions on Pattern Analysis and Machine Intelligence, 19*(4), 302–314. doi:10.1109/34.587996.

Jain, A. K., Nandakumar, K., & Ross, A. (2005). Score normalization in multimodal biometric systems. *Pattern Recognition, 38*, 2270–2285. doi:10.1016/j.patcog.2005.01.012.

Jing, X. Y., Yao, Y. Γ., Yang, J. Y., Li, M., & Zhang, D. (2007). Face and palmprint pixel level fusion and kernel DCV-RBF classifier for small sample biometric recognition. *Pattern Recognition, 40*, 3209–3224. doi:10.1016/j.patcog.2007.01.034.

Kim, S.-W., Kim, K., Lee, J.-H., & Cho, D. (2001). Application of fuzzy logic to vehicle classification algorithm in Loop/Piezo-sensor fusion systems. *Asian Journal of Control, 3*(1), 64–68. doi:10.1111/j.1934-6093.2001.tb00044.x.

Kludas, J., Bruno, E., & Marchand-Maillet, S. (2008). Information fusion in multimedia information retrieval. *Lecture Notes in Computer Science, 4918*, 147–159. doi:10.1007/978-3-540-79860-6_12.

Kokar, M. M., Weyman, J., & Tomasik, J. A. (2004). Formalizing classes of information fusion systems. *Information Fusion, 5*, 189–202. doi:10.1016/j.inffus.2003.11.001.

Kumar, A., & Shekhar, S. (2010). Palmprint recognition using rank level fusion. In *Proceedings of IEEE International Conference on Image Processing*, (pp. 3121-3124). Hong Kong, China: IEEE.

Linas, J., Bowman, C., Rogova, G., Steinberg, A., Waltz, E., & White, F. (2004). Revisiting the JDL data fusion model II. In *Proceedings of 7th International Conference on Information Fusion*. Stockholm, Sweden: IEEE.

Liu, X., & Chen, T. (2003). Geometry-assisted statistical modeling for face mosaicing. In *Proceedings of IEEE International Conference on image Processing*, (vol. 2, pp. 883-886). Barcelona, Spain: IEEE.

Matsumoto, T., Matsumoto, H., Yamada, K., & Hoshino, S. (2002). Impact of artificial 'gummy' fingers on fingerprint systems. []. Springer.]. *Proceedings of SPIE Optical Security and Counterfeit Deterrence Techniques IV, 4677,* 275–289. doi:10.1117/12.462719.

Mitra, S., & Pal, S. K. (2005). Fuzzy sets in pattern recognition and machine intelligence. *Fuzzy Sets and Systems, 156,* 381–386. doi:10.1016/j.fss.2005.05.035.

Monwar, M., Gavrilova, M., & Wang, Y. (2011). A novel fuzzy multimodal information fusion technology for human biometric traits identification. In *Proceedings of ICCI*CC.* Banff, Canada: IEEE.

Moon, Y. S., Yeung, H. W., Chan, K. C., & Chan, S. O. (2004). Template synthesis and image mosaicing for fingerprint registration: An experimental study. In *Proceedings of IEEE International Conference on Acoustics, Speech, and Signal Processing,* (vol. 5, pp. 409-412). Montreal, Canada: IEEE.

Nandakumar, K., Chen, Y., Dass, S. C., & Jain, A. K. (2009). Likelihood ratio-based biometric score fusion. *IEEE Transactions on Pattern Analysis and Machine Intelligence, 30*(2), 342–347. doi:10.1109/TPAMI.2007.70796 PMID:18084063.

Oviatt, S. (2003). Advances in robust multimodal interface design. *IEEE Computer Graphics and Applications, 23*(5), 62–88. doi:10.1109/MCG.2003.1231179.

Pennock, D. M., & Horvitz, E. (2000). Social choice theory and recommender systems: Analysis of the axiomatic foundations of collaborative filtering. In *Proceedings of Seventeenth National Conference on Artificial Intelligence and Twelfth Conference on Innovative Applications of Artificial Intelligence,* (pp. 729-734). Austin, TX: IEEE.

Phillips, P. J., Flynn, P. J., Scruggs, T., Bowyer, K. W., Chang, J., & Hoffman, K. Worek, W. (2005). Overview of the face recognition grand challenge. In *Proceedings of IEEE Computer Society Conference on Computer Vision and Pattern Recognition,* (pp. 947-954). San Diego, CA: IEEE.

Pihur, V., Datta, S., & Datta, S. (2008). Finding cancer genes through meta-analysis of microarray experiments: Rank aggregation via the cross entropy algorithm. *Genomics, 92,* 400–403. doi:10.1016/j.ygeno.2008.05.003 PMID:18565726.

Poh, N., & Bengio, S. (2005). How do correlation and variance of base-experts affect fusion in biometric authentication tasks? *IEEE Transactions on Acoustics, Speech, and Signal Processing, 53,* 4384–4396. doi:10.1109/TSP.2005.857006.

Raghavendra, R., Rao, A., & Kumar, G. H. (2010). Multisensor biometric evidence fusion of face and palmprint for person authentication using particle swarm optimization (PSO). *International Journal of Biometrics, 2*(1), 19–33. doi:10.1504/IJBM.2010.030414.

Rattani, A., Kisku, D. R., Bicego, M., & Tistarelli, M. (2010). Feature level fusion of face and fingerprint biometrics. In *Proceedings of 1st IEEE International Conference on Biometrics: Theory, Applications and Systems.* Washington, DC: IEEE.

Ross, A., & Govindarajan, R. (2005). Feature level fusion using hand and face biometrics. In *Proceedings of SPIE Conference on Biometric Technology for Human Identification II,* (pp. 196-204). Orlando, FL: SPIE.

Ross, A., & Jain, A. K. (2003). Information fusion in biometrics. *Pattern Recognition Letters, 24,* 2115–2125. doi:10.1016/S0167-8655(03)00079-5.

Ross, A., & Jain, A. K. (2004). Multimodal biometrics: An overview. In *Proceedings of 12th European Signal Processing Conference*, (pp. 1221-1224). Vienna, Austria: IEEE.

Ross, A., Nandakumar, K., & Jain, A. K. (2006). *Handbook of multibiometrics*. New York, NY: Springer-Verlag.

Sanderson, C., & Paliwal, K. K. (2001). Information fusion for robust speaker verification. In *Proceedings of Seventh European Conference on Speech Communication and Technology*, (pp. 755-758). Alborg, Denmark: IEEE.

Shafer, G. (1976). *A mathematical theory of evidence*. Princeton, NJ: Princeton University Press.

Sim, T., Baker, S., & Bsat, M. (2003). The CMU pose, illumination, and expression database. *IEEE Transactions on Pattern Analysis and Machine Intelligence, 25*(12), 1615–1618. doi:10.1109/TPAMI.2003.1251154.

Solaiman, B., Pierce, L. E., & Ulaby, F. T. (1999). Multisensor data fusion using fuzzy concepts: Application to land-cover classification using ERS-1/JERS-1 SAR composites. *IEEE Transactions on Geoscience and Remote Sensing, 37*, 1316–1326. doi:10.1109/36.763295.

Soltane, M., Doghmane, N., & Guersi, N. (2010). Face and speech based multi-modal biometric authentication. *International Journal of Advanced Science and Technology, 21*, 41–56.

Truchon, M. (1998). *An extension of the condorcet criterion and kemeny orders. Cahier 9813*. Rennes, France: University of Rennes.

Tumer, K., & Gosh, J. (1999). Linear order statistics combiners for pattern classification. In *Proceedings of Combining Artificial Neural Networks* (pp. 127–162). IEEE.

Wang, C., & Gavrilova, M. (2005). A novel topology-based matching algorithm for fingerprint recognition in the presence of elastic distortions. In *Proceedings of International Conference on Computational Science and its Applications*, (vol. 1, pp. 748-757). Springer.

Wang, Y. (2009). Toward a formal knowledge system theory and its cognitive informatics foundations. *Transactions of Computational Science, 5*, 1–19.

Wang, Y.-P., Dang, J.-W., Li, Q., & Li, S. (2007). Multimodal medical image fusion using fuzzy radial basis function neural networks. In *Proceedings International Conference on Wavelet Analysis and Pattern Recognition*, (vol. 2, pp. 778-782). Beijing, China: IEEE.

Westerveld, T., & de Vries, A. P. (2004). Multimedia retrieval using multiple examples. In *Proceedings of International Conference on Image and Video Retrieval*. IEEE.

Wu, L., Cohen, P. R., & Oviatt, S. L. (2002). From members to team to committee - A robust approach to gestural and multimodal recognition. *IEEE Transactions on Neural Networks, 13*.

Wu, S., & McClean, S. (2006). Performance prediction of data fusion for information retrieval. *Information Processing & Management, 42*, 899–915. doi:10.1016/j.ipm.2005.08.004.

Wu, Y., Chang, K. C.-C., Chang, E. Y., & Smith, J. R. (2004). Optimal multimodal fusion for multimedia data analysis. In *Proceedings of the 12ᵗʰ Annual ACM International Conference on Multimedia,* (pp. 572-579). ACM Press.

Yan, R., & Hauptmann, A. G. (2003). The combination limit in multimedia retrieval. In *Proceedings of the Eleventh ACM International Conference on Multimedia*, (pp. 339-342). ACM Press.

Yu, P., Xu, D., Zhou, H., & Li, H. (2009). Decision fusion for hand biometric authentication. In *Proceedings of IEEE International Conference on Intelligent Computing and Intelligent Systems*, (vol. 4, pp. 486-490). Shanghai, China: IEEE.

Zadeh, L. A. (1965). Fuzzy sets. *Information and Control*, *8*, 338–353. doi:10.1016/S0019-9958(65)90241-X.

Zhao, R., & Grosky, W. I. (2002). Narrowing the semantic gap - Improved text-based web document retrieval using visual features. *IEEE Transactions on Multimedia*, *4*(2), 189–200. doi:10.1109/TMM.2002.1017733.

Chapter 5
Rank Level Fusion

ABSTRACT

Rank level fusion is one of the after matching fusion methods used in multibiometric systems. The problem of rank information aggregation has been raised before in various fields. This chapter extensively discusses the rank level fusion methodology, starting with existing literature from the last decade in different application scenarios. Several approaches of existing biometric rank level fusion methods, such as plurality voting method, highest rank method, Borda count method, logistic regression method, and quality-based rank fusion method, are discussed along with their advantages and disadvantages in the context of the current state-of-the-art in the discipline.

1. INTRODUCTION

Arguably, one of the critical components of the multimodal biometric system development is an information fusion module. It is also a component which is most versatile in the form of input data (processed or unprocessed), types of features (geometric, signal, appearance-based, etc), and decision making process (adaptive, intelligent, fuzzy, learning-based, heuristic-based) it can utilize. Needless to say, the initial choice of bio-

metric—physical, behavioral, soft, or social would both be an input to the information fusion process and dictate some of the choices to be made.

A general rule in theory assumes that the integration of data at an early stage of processing leads to systems which might be more accurate than those where the integration is introduced at later stages. Unfortunately, in practice, fusion at sensor level is hard to achieve, due to the different natures of the biometric traits, which might be hardly compatible (e.g., fingerprint and face).

DOI: 10.4018/978-1-4666-3646-0.ch005

Moreover, most commercial biometric systems do not provide access to the feature sets vanishing the feasibility of a fusion at feature level. Fusions at matching level and at decision level do not require the creation of new databases or matching modules (the ones which constitute the monomodal subsystems are employed).

In general, fusion at matching level can get the job done, but robust and efficient normalization techniques are necessary in the decision module. Normalization technique can be time consuming and selecting inappropriate normalization technique can lead to very poor recognition performance (Ross & Jain, 2003; Gavrilova & Monwar, 2011). Fusion at the decision level performance is hindered due to limited information which is restricted to the Boolean outputs of the subsystems' decision modules, and, in some cases, to the quality scores of the samples. But it is the only possibility of integration at hand if the match scores of the subsystems are not available. Thus, fusion at the rank level is a feasible approach compared to others which consolidates outputs of different classifiers in which no actual matching scores, but only the relative positions of the user/identifier are needed (Monwar & Gavrilova, 2009). Very limited research has been conducted on fusion at this level which has the potential of efficiently consolidating rank information in any multimodal biometric identification system (Gavrilova & Monwar, 2008). In the subsequent section, we will take a look at some of the existing approaches in rank-level fusion.

2. REVIEW OF EXISTING METHODS

The rank level fusion approach is used in biometric identification systems when the individual matcher's output is a ranking of the "candidates" in the template database sorted in a decreasing order of match scores (or, an increasing order of distance score in appropriate cases). The system is expected to assign a higher rank to a template

that is more similar to the query. Plurality voting method, highest rank method, Borda count method, logistic regression method, Bayesian method and quality based method are reported in the literature to perform rank level fusion in multibiometric system (Abaza & Ross, 2009; Monwar & Gavrilova, 2009; Gavrilova & Monwar, 2008; Monwar & Gavrilova, 2010). All of these biometric rank fusion approaches are discussed in the following subsections of this chapter.

The rank information aggregation problem has been addressed in various fields such as (1) in social choice theory which studies voting algorithms which specify winners of elections or winners of competitions in tournaments, (2) in statistics when studying correlation between rankings, (3) in distributed databases when results from different databases must be combined, (4) in collaborative filtering, and (5) in bioinformatics when gene expression similarity search, meta-analysis of microarray data is needed (Truchon, 1998; Fagin, 1999; Pennock & Horvitz, 2000; Pihur, Datta, & Datta, 2008). The criterion for success is the position of the true class in the consensus ranking, as compared to its position in the rankings before fusion.

One of the contributions of rank aggregation research is the work of Farah and Vanderpooten (2008) in which they focused on the rank aggregation problem (also referred to data fusion problem), where rankings of documents, searched into the same collection and provided by multiple methods, are combined in order to produce a new ranking. The authors proposed a new outranking method within a multiple criteria framework using aggregation mechanisms for judging whether a document should get a better rank than another. Their research showed that the proposed method outperforms some other popular classical positional data fusion methods.

Another important contribution in this area is the work of Ailon (2007), where author discussed rank aggregation from partial ranking lists. One of the main drawbacks of considering full rankings is that obtaining full ranking information from all sources is not always possible. Authors state that it is unlikely that a search engine would give a ranking of the entire set of all web pages for a given query. Instead, only the first few top-ranked pages are returned. Authors note that partial rankings arise naturally in many everyday problems: in many sports tournaments, ties are possible outcomes of single matches, while in an election system, each voter can tie together a subset of candidates as a way of expressing neutrality with respect to that subset (Ailon, 2007). The author reported two approximation algorithms for aggregating partial rankings. The first is a new 2-approximation, generalizing a well-known 2-approximation for full rank aggregation. The second approximation algorithm is a new 3/2-approximation algorithm, generalizing a recent algorithm (Ailon, Charikar, & Newman, 2005) for full rank aggregation to the problem of partial rank aggregation.

In the area of bioinformatics, one important rank information fusion research is the work of Pihur et al. from University of Louisville, USA (Pihur, Datta, & Datta, 2008). In their research, they used a weighted rank aggregation method based on the Cross-entropy Monte Carlo algorithm to find cancer genes through meta-analysis of microarray experiments. Their proposed meta-analysis approach to microarray data was a two-step procedure. The first step was individual analysis, where by analyzing each microarray dataset individually, a set of "interesting" genes (top-k) that exhibit the largest differences in terms of expression values between the groups was obtained for each dataset. The second step was rank aggregation, where the individual lists obtained from the first step on the rankings of genes within each list is performed to produce a "super"-list of k genes which would reflect the overall importance of genes as judged by the collective evidence of all experiments.

Very recently, some pattern recognition researchers have started to investigate rank level fusion in the context of multimodal biometrics. In 2009, Abaza and Ross suggested several modifications to the highest rank and Borda count methods to enhance the performance of a quality based rank-level fusion scheme in the presence of weak classifiers or low quality input images (Abaza & Ross, 2009). Their experiments conducted on a multimodal database consisting of a few hundred users demonstrated that the suggested modifications enhance the rank-1 accuracy. Further, their experiments also indicated that including image quality in the fusion scheme can enhance the Borda count rank-1 accuracy (Abaza & Ross, 2009).

In another attempt in 2010, Kumar and Shekhar (2010) investigated a new approach for personal recognition using rank level combination of multiple palmprint representations. They used Borda count, weighted Borda count, maximum rank, and nonlinear weighted ranking method on two palmprint image databases. Among all of the fusion approaches they investigated, the usage of nonlinearities in conjunction with the weights resulted in the highest performance improvement.

We now will take a look at specific algorithms pertaining to most popular rank fusion methods, namely Plurality voting, Highest rank, Borda count, Logistic regression and Quality-based rank fusion methods.

3. PLURALITY VOTING RANK FUSION METHOD

The plurality voting method is a positional method for rank aggregation which takes into account information about individuals' preference orderings (Abaza & Ross, 2009). However, this method does not take into account a matcher's entire preference ordering, instead uses only information about each voter's most preferred alternative. This method is good for combining a small number of specialized matchers. In this method, the consensus ranking is obtained by sorting the identities according to their number of position in the top position. The algorithm is adopted from Abaza and Ross (2009) (See Algorithm 1).

Algorithm 1: Plurality voting
Step 1: Get three ranking lists from different biometric classifiers.
Step 2: For all ranking lists -
    Step 2a: Find out the identity which appears most at the top of the three ranking list.
    Step 2b: If any alternative is found then position that identity to the available position in the consensus ranking list starting from the top.
    Step 2c: If no alternative is found, then find the identity which has the highest rank for that position and put that identity to the available position in the consensus ranking list starting from the top.
    Step 2d: Go to step 2 and start from the next position.

For example, in a five classifiers system, suppose user1 is chosen as the top ranked identity by classifier 1 and classifier 3. User 2 is chosen as top ranked identity by classifier 2, user 5 is chosen as top ranked identity by classifier 4 and user 20 is chosen as top ranked identity by classifier 5. Then according to the plurality voting rank fusion method, user 1 will be chosen as top ranked identity in the consensus ranking list.

The advantage of the plurality voting rank fusion method is that this method can overcome the unwanted behavior of any classifier. Suppose a weak classifier chose an identity as the top ranked identity, but the identity is not supposed to be at the top ranked position. If the other classifiers decide not to choose the identity at the top position, that identity will not be at the top position in the consensus ranking list obtained by the plurality voting rank fusion method. The problem of this method is that, in this method, only the top position in any initial ranking list is considered, which frequently results in an un-reliable decision from the multibiometric system.

4. HIGHEST RANK METHOD FOR RANK FUSION

The highest rank method is good for combining a small number of specialized matchers and hence can be effectively used for a multimodal biometric system where the individual matchers perform well. In this method, the consensus ranking is obtained by sorting the identities according to their highest rank.

The steps in Algorithm 2 show the procedure of employing highest rank fusion method in a multimodal biometric system.

Algorithm 2: Highest rank
Step 1: Get the ranking lists from different biometric classifiers.
Step 2: For all ranking lists -
    Step 2a: For all identities in the three ranking lists -
    Step 2a(i): Find out the consensus rank of each identity
utilizing the following equation –
Consensus rank,

$$R_c = \min_{i=1}^{n} R_i \qquad (5.1)$$

where, n is the number of ranking lists.

Step 3: Sort R_c in ascending order and replace with corresponding identity.

The advantage of this method is the ability to utilize the strength of each matcher. Even if only one matcher assigns the highest rank to the correct user, it is still very likely that the correct user will receive the highest rank after reordering. The disadvantage of this method is that the final ranking may have many ties (Ho, Hull, & Srihari, 1994; Monwar & Gavrilova, 2009). Usually, ties are broken randomly, which sometimes may lead to the accepting incorrect decision from the weakest classifier (Ho, Hull, & Srihari, 1994). Another disadvantage of this method is that, similar to plurality voting rank fusion method, in this method only the top position in any initial ranking list is considered, which frequently results in an un-reliable decision from the multibiometric system. The number of classes sharing the same ranks depends on the number of classifiers used. Due to this property, this method is not the best choice for a security critical multimodal biometric system.

Recently, Abaza and Ross (2009) developed a modification to the equation of the existing highest rank method to solve the tie problem by introducing a perturbation factor with the using the Borda count method. According to their modification, the consensus rank of a particular class is obtained by (Abaza & Ross, 2009):

Consensus rank,

$$R_c = \min_{i=1}^{m} R_i + \varepsilon \qquad (5.2)$$

where,

$$\varepsilon = \frac{\sum_{i=1}^{m} R_i}{K} \qquad (5.3)$$

Here, K is a large value used to ensure that ε remains small. Their rationale was to consider a perturbation term which biases the fused rank by considering *all* the ranks associated with a particular user (Abaza & Ross, 2009). For example, in the case of fusing outputs of two classifiers, assume that the rank for one user by the first classifier is 1 and by the second classifier is 2. Similarly, for another user, suppose that the rank by

the first classifier is 3 and by the second classifier is 1. So, the consensus ranks for both the user would be 1 according to the highest rank method. In this case, a tie occurs. But according to the equation 5.2, the consensus rank of the first user is *(1+3/100)* or 1.03 and the consensus rank of the second user is *(1+4/100)* or 1.04, where K = 100 in equation 5.3. So, in the consensus ranking list, the first user is ranked higher than the second user. Hence the ties are broken.

5. BORDA COUNT RANK FUSION METHOD

Borda count rank fusion method, proposed in 1771 by the French mathematician Jean-Charles de Borda (1781), is a procedure in which each classifier forms a preference ranking for all identities. The Borda count method is the most widely used rank aggregation method and uses the sum of the ranks assigned by individual matchers to calculate the final rank.

Algorithm 3 shows the Borda score calculation process in this work (Borda, 1781).

Algorithm 3: Borda count
Step 1: Get the ranking lists from different biometric classifiers.
Step 2: For all ranking lists –
    Step 2a: For all identities in the three ranking list
        Step 2a(i): Find out the total Borda score of each identity utilizing the following equation –
Total Borda score,

$$B_c = \sum_{i=1}^{n} B_i \qquad (5.4)$$

where, n is the number of ranking list and B_i is the Borda score in the i-th ranking list.

Step 3: Sort B_c in descending order and replace with corresponding identity.

This method works under assumption that that the ranks assigned are independent and the quality of matchers are similar. The Borda count for each class represents the consensus of the matchers that the input pattern belongs to that class. The advantage of this method is that it is easy to implement and requires no training stage. These properties made the Borda count method feasible to incorporate in multimodal biometric systems. The disadvantage of this method is that it does not take into account the differences in the individual matcher's capabilities and assumes that all the matchers perform equally well, which is usually not the case in most real biometric systems. This makes the Borda count method highly vulnerable to the effect of weak classifiers. For example, suppose there are 5 classifiers. Assume that for an identity, 4 of the 5 classifiers result in rank 1 while the *5th* results in rank 27: so the Borda score for that user is 31. Assume that for another user, the ranks assigned by the classifiers are 2, 2, 3, 6, 2, and hence the Borda score is 15. So, in the consensus ranking list, the second person will be ranked higher as his/her Borda score is lower than that of the first user. This is due to the weak performance of just one classifier. To overcome this problem, in 2009, Abaza and Ross (2009) suggested a modification of the existing Borda count method to improve the performance by discarding the outcomes of the classifier(s) which is not able to produce good results. According to their modifications, the worst ranks are eliminated before invoking the fusion scheme based on Nanson function (Fishburn, 1990), which can be termed as Borda elimination.

Nanson function is to first eliminate the weakest rank, i.e.,

$$\max_{i=1}^{m} B_i = 0 \qquad (5.5)$$

and then compute the regular Borda count on the remaining ranks (Abaza & Ross, 2009). In this implementation, the weakest rank is therefore eliminated. For the previous example, by applying the Nanson modification, the rank for the 5th classifier (weakest) will be 0. Then, the fused Borda score of the first user will be 4 and the Borda score of the second user will be 9. Thus, the first user will be selected above the second user in the final consensus ranking list.

6. LOGISTIC REGRESSION RANK FUSION METHOD

The logistic regression method, which is a variation of the Borda count method, calculates the weighted sum of the individual ranks (Ho, Hull, & Srihari, 1994). In this method, the final consensus rank is obtained by sorting the identities according to the sum of their ranks obtained from individual matchers multiplied by the weights.

Algorithm 5 shows the Borda score calculation process in this work according to Ho, Hull, and Srihari (1994).

Algorithm 5: Logistic regression
Step 1: Get the ranking lists from different biometric classifiers.
Step 2: Assign different weights to all ranking lists.
Step 3: For all ranking lists –
    Step 3a: For all identities in the three ranking list
        Step 3a(i): Find out the total Borda score of each identity utilizing the following equation –
Total Borda score,

$$R_c = \sum_{i=1}^{m} W_i R_i \qquad (5.6)$$

where, n is the number of ranking list, R_i is the Borda score in the i-th ranking list and W_i is the weight assigned to the i-th classifier.

Step 4: Sort R_c in descending order and replace with corresponding identity.

The weight to be assigned to the different matchers is determined by the recognition performances obtained through numerous trial executions of the system and through applying common knowledge. This method is very useful when the different matchers have significant differences in their accuracies, but requires a training phase to determine the weights. Also, one of the key factors that have direct effect on the performance of a biometric system is the quality of the biometric samples. Hence, the single matchers' performance can vary with different sample sets which make the weights allocating process more challenging and inappropriate weight allocation can eventually reduce the recognition performance of this multimodal biometric system (using logistic regression) compared to unimodal matchers. So, in some cases, logistic regression method cannot be employed for rank aggregation.

Figure 1 illustrates the normal highest rank, the normal Borda count and the logistic regression rank fusion approaches. In this figure, the less the value of the rank, the more accurate the result is. Here, the rank for 'person 1' is 1, 2, and 2, respectively, from the face, ear and signature matchers. For the highest rank method, the fused score is 1 for person 1. Similarly, for person 2, person 3, person 4, and person 5, the fused ranks are 1, 3, 2, and 3, respectively. There is a tie between person 1 and person 2 and 'person 3' and 'person 5'. These ties are broken arbitrarily. So, in the final reordered ranking, 'person 1' gets the top position in the reordered rank list whereas, 'person 2' is in the second position.

For the Borda count method, the initial ranks are first added. Thus, 5, 7, 13, 9, and 11 can be found as the fused score for 'Person 1' to 'Person 5' respectively. So, 'Person 1' gets the top position in the reordered list due to his/her lowest fused score, 'Person 2' gets the second position and so on.

For the logistic regression method, the matchers need to be assigned weights which are determined by the recognition performance of the matchers. For this system, suppose face matcher is assigned a weight of 0.1, ear matcher is assigned a weight of 0.5 and signature matcher is assigned a weight of 0.4. This weight assignment can be done by evaluating the performance of the three matchers with a number of experiments and by researching the previous investigations of these matchers. For this system, it is assumed the matcher with the minimum weight works better than the other matchers. So, face matcher works better than the ear matcher or signature matcher. The fused scores for different identities are calculated by multiplying their positions in the initial rank lists with the appropriate weight assigned to each matcher. Thus, 2.2, 1.4, 4.8, 3.4, and 3.5 fused scores are found for 'Person 1' to 'Person 5' respectively. So, 'Person 2' is on top position in the reordered ranking list.

7. QUALITY-BASED RANK FUSION METHOD

Quality-based rank fusion method depends not only on the ranking list of the unimodal classifiers, but also on the quality of the input images. Usually, this method applies on other biometric rank fusion approach with the modification by incorporating the quality of the input image. Quality based fusion methods usually do not have any training phase and hence can be used in other biometric information fusion process, such

Figure 1. Example of rank level fusion (adopted from Ross, Nandakumar, & Jain, 2006)

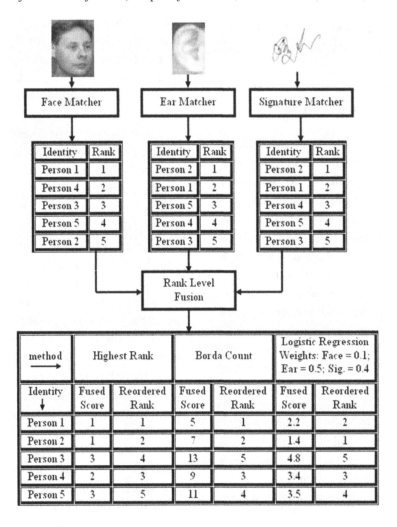

as fuzzy logic based fusion. There is no specific rule or general equation for quality based fusion method. Researchers can apply this method to any of their existing methods to improve the identification or verification rate. For example, Abaza and Ross introduced a quality based rank fusion method by modifying the existing Borda count method incorporating the quality of the input image into the equation (Abaza & Ross, 2009). Figure 2 show a sample block diagram of the quality based rank fusion method.

As stated earlier, the main drawback of the Borda count method is its inability to account for one or more weak classifiers. This was the moti-

vation behind adding statistically calculated weights to different classifier outputs as in the logistic regression method. However, computation of these weights needs a training phase for different classifiers. But when the image quality can be incorporated into the process, then no extra training phase will be required. In this process, the Borda count method will directly make use of the input data quality.

Thus the steps for calculating the Borda score of one user is redefined as Algorithm 6.

Algorithm 6: Quality-based fusion

Figure 2. Quality-based rank fusion (adopted from Abaza & Ross, 2009)

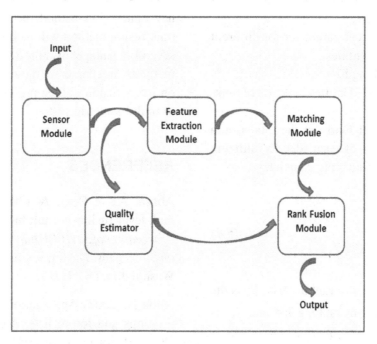

Figure 3. Good and bad quality input sample of a biometric system

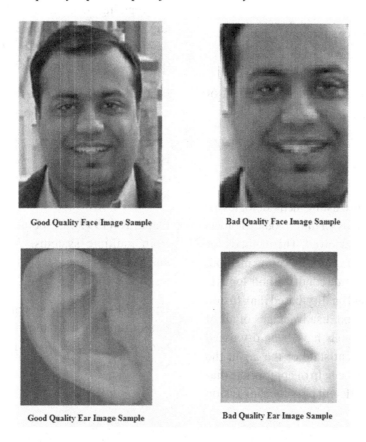

Step 1: Get the ranking lists from different biometric classifiers.

Step 2: Assign different parameter for different quality to all identities.

Step 3: For all ranking lists –

Step 3a: For all identities in the three ranking list

Step 3a(i): Find out the total Borda score of each identity utilizing the following equation –

Total Borda score,

$$B_c = \sum_{i=1}^{n} Q_i B_i \qquad (5.7)$$

where, n is the number of ranking list, B_c is the Borda score in the i-th ranking list and Q_i is defined as $Q_i = \min(Q_i)$ for that particular user and represent the quality of the probe and gallery fingerprint impressions.

Step 4: Sort B_c in descending order and replace with corresponding identity.

The weight factor, Q_i, reduces the effect of poor quality biometric samples (See Figure 3).

8. SUMMARY

In this chapter, different existing methodologies for rank level fusion methods for multimodal, biometric system have been reviewed. The methods for rank level fusion include plurality voting method, highest rank method, Borda count method, logistic regression method, and quality-based rank fusion method. Advantages and disadvantages of all of these rank fusion methods have been discussed in the context of current state of the art in the discipline. Also, with the help of appropriate diagrams, outcomes of different possible rank fusion methods have been shown. In the next chapter, a new rank fusion method, the Markov chain based rank fusion method will be discussed which has several advantages over the traditional rank fusion methods. In chapter 7, the new approach based on fuzzy fusion will be presented in the context of multibiometric system.

REFERENCES

Abaza, A., & Ross, A. (2009). Quality based rank-level fusion in multibiometric systems. In *Proceedings of 3rd IEEE International Conference on Biometrics: Theory, Applications and Systems*. Washington, DC: IEEE.

Ailon, N. (2007). Aggregation of partial rankings, p-ratings and top-m lists. *Algorithmica*, *57*(2), 284–300. doi:10.1007/s00453-008-9211-1.

Ailon, N., Charikar, M., & Newman, A. (2005). Aggregating inconsistent information: Ranking and clustering. In *Proceedings of 37th Annual ACM Symposium on Theory of Computing (STOC)*, (pp. 684-693). Baltimore, MD: ACM.

Black, D. (1963). *The theory of committees and elections* (2nd ed.). Cambridge, UK: Cambridge University Press.

Borda, J. C. (1781). *M'emoire sur les 'elections au scrutin*. Paris, France: Histoire de l'Acad'emie Royale des Sciences.

Fagin, R. (1999). Combining fuzzy information from multiple systems. *Journal of Computer and System Sciences*, *58*(1), 83–99. doi:10.1006/jcss.1998.1600.

Farah, M., & Vanderpooten, D. (2008). An outranking approach for information retrieval. *Information Retrieval*, *11*(4), 315–334. doi:10.1007/s10791-008-9046-z.

Fishburn, P. C. (1990). A note on "A Note on Nanson's Rule". *Public Choice*, *64*(1), 101–102. doi:10.1007/BF00125920.

Gavrilova, M., & Monwar, M. M. (2008). Fusing multiple matcher's outputs for secure human identification. *International Journal of Biometrics*, *1*(3), 329–348. doi:10.1504/IJBM.2009.024277.

Gavrilova, M., & Monwar, M. M. (2011). Current trends in multimodal system development: Rank level fusion. In Wang, P. (Ed.), *Pattern Recognition, Machine Intelligence and Biometrics (PRMIB)*. Berlin, Germany: Springer. doi:10.1007/978-3-642-22407-2_25.

Ho, T. K., Hull, J. J., & Srihari, S. N. (1994). Decision combination in multiple classifier systems. *IEEE Transactions on Pattern Analysis and Machine Intelligence*, *16*(1), 66–75. doi:10.1109/34.273716.

Kumar, A., & Shekhar, S. (2010). Palmprint recognition using rank level fusion. In *Proceedings of IEEE International Conference on Image Processing*, (pp. 3121-3124). Hong Kong, China: IEEE.

Monwar, M. M., & Gavrilova, M. (2009). A multimodal biometric system using rank level fusion approach. *IEEE Transactions on Systems, Man, and Cybernetics. Part B, Cybernetics*, *39*(4), 867–878. doi:10.1109/TSMCB.2008.2009071 PMID:19336340.

Monwar, M. M., & Gavrilova, M. (2010). Secured access control through Markov chain based rank level fusion method. In *Proceedings of the 5th International Conference on Computer Vision Theory and Applications (VISAPP)*, (pp. 458-463). Angers, France: VISAPP.

Pennock, D. M., & Horvitz, E. (2000). Social choice theory and recommender systems: Analysis of the axiomatic foundations of collaborative filtering. In *Proceedings of Seventeenth National Conference on Artificial Intelligence and Twelfth Conference on Innovative Applications of Artificial Intelligence*, (pp. 729-734). Austin, TX: IEEE.

Pihur, V., Datta, S., & Datta, S. (2008). Finding cancer genes through meta-analysis of microarray experiments: Rank aggregation via the cross entropy algorithm. *Genomics*, *92*, 400–403. doi:10.1016/j.ygeno.2008.05.003 PMID:18565726.

Ross, A., & Jain, A. K. (2003). Information fusion in biometrics. *Pattern Recognition Letters*, *24*, 2115–2125. doi:10.1016/S0167-8655(03)00079-5.

Ross, A., Nandakumar, K., & Jain, A. K. (2006). Handbook of multibiometrics. New York, NY: Springer-Verlag.

Truchon, M. (1998). *An extension of the Condorcet criterion and Kemeny orders. Cahier 9813*. Rennes, France: University of Rennes.

Chapter 6
Markov Chain for Multimodal Biometric Rank Fusion

ABSTRACT

Markov chain is a mathematical model used to represent a stochastic process. In this chapter, Markov chain-based rank level fusion method for multimodal biometric authentication system is discussed. Due to some inherent problems associated with existing biometric rank fusion methods, Markov chain-based biometric rank fusion has recently emerged in biometric context. The notion of Markov chain and its construction mechanisms are presented along with discussion on some early research conducted on Markov chain in other rank aggregation frameworks. This chapter also presents a detailed description of recent experimentations conducted to evaluate the performance of Markov chain-based biometric rank fusion method in a face, ear, and iris-based application framework.

1. INTRODUCTION

In the previous chapter, different methodologies for rank level fusion were presented. These methods included plurality voting method, highest rank method, Borda count method, logistic regression method and quality based rank fusion method for multimodal biometric system. Among those methods, the logistic regression method consistently provides high Performance, however it still has some drawbacks. The results obtained through this method can be varied significantly for different datasets due to their diverse qualities. Logistic regression method for a multimodal dataset with the same image quality will produce results similar to Borda count method, as the assigned weights to different biometric matchers' outputs will be the same. Thus, allocating appropriate weights to different matchers (comparing different quality datasets) requires appropriate learning technique,

DOI: 10.4018/978-1-4666-3646-0.ch006

which is time consuming. Also, inappropriate weight allocation can result in wrong recognition results. Further, the size of the multimodal biometric database is usually large and thus only the top few results are considered for the final reordered ranking. Hence, a very common scenario of a rank based multimodal biometric system is that some results may rank at top by a few classifiers and the rest of the classifiers do not even output the result. In this situation, the logistic regression approach cannot produce a good recognition performance. Thus, a novel rank fusion method utilizing Markov chain has been recently developed at BT Lab at the University of Calgary. This method can be efficiently used in multimodal biometric authentication system comprised of varied quality datasets. The method has been successfully used in other information fusion applications. In this chapter, an overall description of this method is given. It includes Markov chain definition, advantages and disadvantages of Markov chain in multimodal biometric fusion scenario, previous research on Markov chain and its application in rank level fusion.

2. MARKOV CHAIN

A Markov chain is named for the Russian mathematician Andrei Andreyevich Markov. It is a mathematical model that can be thought of a being in exactly one of a number of states at any time (Markov, 1906). A Markov chain has a set of *states*, $S = \{s_1; s_2; ::: ; s_r\}$. The process starts in one of these states and moves successively from one state to another (Kemeny, Snell, & Thompson, 1974). Each move is referred to as a *step*. If the chain is currently in state s_i, then it can move to state s_j with a probability p_{ij}. This probability is preset at the beginning of the process and does not depend on how the state was reached. The probability p_{ij} is *referred to as transition prob-*

abilities. The process can remain in the same state with probability *pii*. The starting state is given by an initial probability distribution (Kemeny, Snell, & Thompson, 1974).

The following example illustrates how Markov chain operates. Assume that there is a sports team which performance highly depends on its previous history of winning or losing. If the team wins, then there is 50% chance it will win the next game, and 25% chance it will tie or lose the next game. If the team ties, there is 75% it will tie again, and 25% it will lose. Finally, if the team loses, there is 50% chance it will lose next game and 50% chance it will win.

Now we can build a Markov chain. States in this example are W (Win), T(Tie) and L(Lose). Transitional probabilities can be represented in a matrix:

$$P = \begin{matrix} & W & T & L \\ W \\ T \\ L \end{matrix} \begin{pmatrix} 1/2 & 1/4 & 1/4 \\ 0 & 3/4 & 1/4 \\ 1/2 & 0 & 1/2 \end{pmatrix} \quad (6.1)$$

The entries in the first row of the matrix **P** in this example represent the probabilities for the win, tie or lose of the team during the next game. The entries in the second and the third row represent probabilities of win, lose or tie following the tie (second row) or the loss (third row) Such an array is commonly called the *matrix of transition probabilities*, or the *transition matrix*.

The matrix allows to determine, given the state i, the probability of win, loss or tie in one, two, or any number of consequent games in the future.

Let us consider one more detailed example. It showcases the main principle of a Markov chain. Suppose, a library book is shared by two friends, Mike and Charles. If the book is borrowed by Mike during a week, there is an 80% chance that he will keep the book for the next week. On the other hand, if the book is borrowed by Charles, there is an 60% chance that he will keep the book for the next week. With this information, a Markov chain can be formed as follows to solve the question – what percentage of the time do each friend have the book.

States can be taken as the probability of keeping the book by Mike (M) and Charles (C) and from the above information the transition probabilities can be determined.

$$P = \begin{matrix} & M & C \\ M & \\ C & \end{matrix} \begin{bmatrix} .8 & .2 \\ .4 & .6 \end{bmatrix} \tag{6.2}$$

The entries in the first row of the transition matrix **P** in this example represent the probabilities that what will happen in the following week if the book is being kept by Mike. Similarly, the entries in the second row represent the probabilities for the status of the book in the following week if the book is being kept by Charles.

The question of determining the probability is considered that, given the book is being kept in state i this week, it will be in state j two weeks from now. This probability is denoted by p^2_{ij}. Also in this example, it is seen that if the book is being kept by Mike this week, then the event that the book is being kept by Charles two weeks from now is the disjoint union of the following two events:

1. The book is being kept by Mike next week and by Charles two weeks from now,
2. The book is being kept by Charles next week and by Charles two weeks from now.

The probability of the first of these events is the product of the conditional probability that the book is being kept by Mike next week, given that the book is being kept by Mike this week, and the conditional probability that the book is being kept by Charles two weeks from now, given that the

book is being kept by Mike next week (similar to the 'Wizard of Oz' problem described in Kemeny, Snell, & Thompson, 1974).

Using the transition matrix **P**, it can be written that this product is $p_{11}p_{12}$. The other two events also have probabilities that can be written as products of entries of **P**. Thus,

$$p^2_{12} = p_{11}p_{12} + p_{12}p_{22} \tag{6.3}$$

In general, if a Markov chain has r states, then

$$p^2_{ij} = \sum_{k=1}^{r} p_{ik}p_{kj} \tag{6.4}$$

So, we have the transition probability of P^2, which is the probability of the status of the book two weeks from now.

$$P^2 = \begin{matrix} & M & C \\ M & \\ C & \end{matrix} \begin{bmatrix} .72 & .28 \\ .56 & .44 \end{bmatrix} \tag{6.5}$$

In this way, we can get P^6, which is the probability of the status of the book six weeks from now.

$$P^6 = \begin{matrix} & M & C \\ M & \\ C & \end{matrix} \begin{bmatrix} .668 & .332 \\ .664 & .336 \end{bmatrix} \tag{6.6}$$

Or, we can write that,

$$P^6 \approx \begin{matrix} & M & C \\ M & \\ C & \end{matrix} \begin{bmatrix} 2/3 & 1/3 \\ 2/3 & 1/3 \end{bmatrix} \tag{6.7}$$

As, *n* gets larger, P^n gets closer to the following matrix:

$$\begin{bmatrix} 2/3 & 1/3 \\ 2/3 & 1/3 \end{bmatrix}$$

That means, no matter who holds the book for this week (starting week), it can be concluded that the probability that book is being kept by Mike is .67 (2/3) and the probability that book is being kept by Charles is .33 (1/3).

Thus the matrix,

[2/3 1/3]

is called the *stationary matrix* of this Markov chain as it remains stationary for any starting probability distribution. This stationary distribution property of Markov chain can be used in rank aggregation of Markov chain. After constructing a Markov chain by the initial ranking lists (obtained from different classifiers' outputs), a stationary distribution is obtained from which the consensus ranking list is constructed.

3. RESEARCH ON MARKOV CHAIN

Markov chains are applied in a number of ways to many different fields. They can be either used as a mathematical model corresponding to some random physical process, or to simulate an abstract theoretical concept. Application areas of Markov chain include physics (thermodynamics, statistical mechanics), chemistry (enzyme activity), the growth of copolymers, statistics (statistical testing, Bayesian inference, etc.), Internet applications (page rank, analyzing Web navigation behavior of users, etc.), economics, finance, information sciences (Hidden Markov Model for pattern recognition, Viterbi algorithm for error correction), bioinformatics, social sciences education, stock market predictions, music, and sports (Grinstead & Snell, 1997).

In an attempt to select consensus biomarkers form high throughput experimental data, in 2007, Dutkowski and Gambin (2007) used the Markov chain method. They presented two solutions for the consensus biomarker feature selection, with the focus on estimating which criteria leads to the better selection.

The first proposed approach was based on computing a consensus ranking from the outcomes of several feature selection procedures. The stationary distribution of an appropriately defined Markov chain is used in this approach. The states of the chain proposed by the authors correspond to the features ranked by various scoring functions and the transition probabilities depend on the position of the features in the given partial rankings. The aggregated consensus ranking is obtained as the list of states sorted by their stationary probabilities. The author also claims a high performance of the method on large databases due to original approximation algorithm (Dutkowski & Gambin, 2007). The second method proposed was based on Principal Component Analysis (PCA), which is a well known and extensively sued in biometric projection method of original variables with maximal variance. The method worked very well for the given problem as there is a large diversity of the samples. Thus, the authors instead of preserving the overall variance, aimed to retain the variance between classes. In order to increase the discrimination power of the method, authors applied PCA only to the group of the most discriminative variables (Dutkowski & Gambin, 2007).

Another quite different area of application of Markov chain method is for ranking web pages. In Dwork, Kumar, Naor, and Sivakumar (2001), the authors proposed a rank aggregation process to reduce search engine spam in metasearch. They noted that the Kemeny optimal aggregation works well for their purpose, and developed a natural relaxation which they called *local Kemeny optimality*. In their research, they showed how to produce a maximally consistent locally Kemeny optimal solution from any initial ranking. They compared Markov chain method with other positional rank fusion method and found that Markov chain method has some advantages over other rank fusion methods.

In Gambin and Pokarowski (2001), the authors proposed a combinatorial aggregation algorithm for stationary distribution of a large Markov chain. They found that their proposed algorithm performs well when the state space of Markov chain is large enough and when other direct and iterative methods are inefficient. Their method was based on grouping of the states of a Markov chain in such a way that the probability of changing the state inside the group is higher than probability of interactions between groups. Authors claimed that their proposed method can be seen as "an algorithmization of famous Markov Chain Tree Theorem" (Gambin & Pokarowski, 2001). They carried out experimentations of their method on several benchmark examples and showed that the proposed algorithm can be useful in many real world problems.

Sculley (2006) made a different attempt of utilizing Markov chain for rank aggregation of similar items. The author assumed the rank aggregation process as the unsupervised analog to regression, in which the goal is to find an aggregate ranking that minimizes the distance to each of the given ranked lists. In this research, the author addressed the problem of aggregating noisy, incomplete, or disjoint ranked lists through the different way of adding similarity information. Their motivation for this approach was that similar items should be ranked similarly, resulting in an appropriate similarity measure for the data.

The author showed several examples where existing standard methods for rank aggregation have been extended in order to include the role of similarity between items. As an example, il-lustrated in this research, the following problem is considered (Sculley, 2006):

There are ranked lists from two experts:

Expert 1: A, B, C
Expert 2: C', D, E

The items C and C' are very similar. While majority of the methods would consider those two lists as different, with results of rank aggregation being something like:

Aggregation 1: A, C', B, D, C, E or
Aggregation 2: C', A, D, B, E, C

According to the author (Sculley, 2006), each of these aggregations is not correct, and he proposes to sue similarity measure to arrive to the following aggregation with similarity between C and C' accounted for:

Aggregation 3: A, B, C', C, D, E

The author (Sculley, 2006) also proposed an extension to the Kendall Tau metric that formalizes the benefits of considering similarity information in ranking lists. He computed the similarity transitions of Markov chain which, similar to many situations, fails when the input lists are disjoint (Figure 1). In the proposed technique, the author introduced similarity transitions which can connect Markov Chain islands, left disjoined in other methods.

The key contribution is that the similarity transition is defined based on the similarity measured between nodes. Thus, the ranking of an item will depend not only on those items it is ranked higher or lower than, but also on those items that are similar to a given item. The author also makes a compelling case for using the proposed method by both advertisers and search engine developers.

Figure 1. Rank aggregation of different items with similarity using Markov chain (adopted from Sculley, 2006)

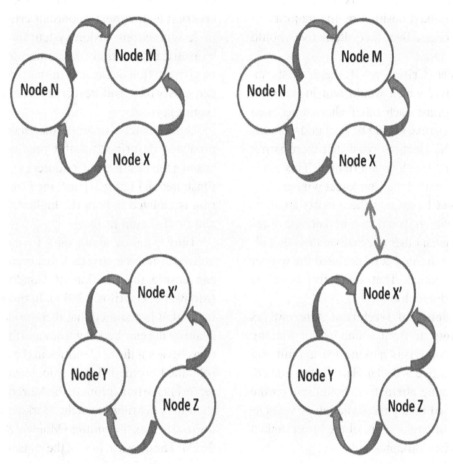

4. MARKOV CHAIN FOR MULTIMODAL BIOMETRIC FUSION

In 2011, Monwar and Gavrilova (2011) utilized Markov chain as a method for biometric rank fusion. This approach brought a new dimension to the current ways of biometric rank aggregation and can be effectively used by the homeland and border security forces and by other intelligence services.

In their research (Monwar & Gavrilova, 2011), they considered the biometric rank aggregation similar to a voting mechanism. In the multimodal biometric rank fusion process, the classifiers are considered as voters. So, if there are three biometric traits used in a multimodal biometric system, the number of voters in the system would be three. Those three voters or classifiers produce three ranking list based on the similarity or distance scores of test and template biometric data. The final process is to combine the ranking lists obtained from three classifiers or voters to make a consensus ranking lists to find out the desired identity or alternative from the system.

In a general voting method evaluation, the most important thing is to ensure the fairness of the voting system. The following are the conventional fairness criteria proposed by a French mathematician in 1800 and adopted in modern voting system research (Condorcet, 1785):

- **Majority Criterion:** If there exists an alternative in the ranking lists, which receives the top ranked position by the majority of the matchers, then that alternative should be the winner.

- **Condorcet Criterion:** If there exists an alternative, which would win in pairwise votes against each other alternative, then that alternative should be declared the winner of the election. Note that there is not necessarily such an alternative. This alternative is called the Condorcet winner.

- **Condorcet Looser:** If there exists an alternative, which would lose in pairwise votes against each other alternative, then that alternative should not be declared the winner of the election. That alternative is called the Condorcet looser.

- **Independence of Irrelevant Alternatives Criterion:** If from a number of ranking lists, a consensus ranking list is built and an alternative is declared as the top ranked, this winning alternative should remain the top-ranked in any recalculation of votes as a result of one or more of the lower ranked alternatives dropping out.

- **Pareto Criterion:** If there is at least one alternative (say alternative a) that every matcher prefers to another (say alternative b) then it should be impossible for b to win.

- **Monotonicity Criterion:** It should be impossible for a top ranked alternative to lose in a re-aggregation process if the only changes in the ranked alternatives are changes that were favorable to that candidate.

- **Neutrality Criterion:** All matchers should be treated equally. No matcher has special influence on any of the alternatives. Similarly, all alternatives should be treated equally. No alternative has more privilege than any other.

Among all of the above-mentioned fairness criteria, researchers considered the Condorcet criterion as the most important criterion. Thus, in designing rank fusion system for multimodal biometric information, more concentration should be given to find out an appropriate method which can satisfy the Condorcet criterion of fair ranking (voting) process.

Unfortunately, none of the rank fusion approaches described in the previous chapters ensures the election of Condorcet winner. This is illustrated in Figure 2. Here, the Condorcet criterion is violated in both the highest rank method and Borda count method.

Thus, Markov chain rank fusion method is utilized in the biometric rank fusion process, which can ensure the election of Condorcet winner (Monwar & Gavrilova, 2011). In this rank fusion method, it is assumed that there exists a Markov chain on the enrolled identities and the order relations between those identities in the ranking lists (obtained from different biometric matchers) represent the transitions in the Markov chain. The stationary distribution of the Markov chain is then utilized to rank the entities (Monwar & Gavrilova, 2011). The construction of the consensus ranking list from the Markov chain method is done according to algorithm developed in (Sculley, 2006) and is summarized in Figure 3.

This Markov chain approach for biometric rank aggregation has several advantages. This method handles the partial ranking list very well and provides a more objective comparison of all candidates against each other.

The Markov chain method also handles the uneven comparison, i.e., when the results of the initial ranking lists are very much different. Thus, in a theoretical work devoted to using Markov chain in rank aggregation scenario (Dwork, Kumar, Naor, & Sivakumar, 2001), authors noted that heuristics for combining rankings are motivated by some underlying principle and the Markov chain model can be viewed as the natural exten-

Figure 2. Highest rank and Borda count rank fusion methods (in both fusion methods, Condorcet criterion is violated)

Face Matcher	Ear Matcher	Iris Matcher
Identity X	Identity Y	Identity M
Identity Y	Identity Z	Identity X
Identity N	Identity N	Identity Y
Identity Z	Identity X	Identity N
Identity M	Identity M	Identity Z

Highest Rank	Borda Count
Identity M	Identity Y
Identity Y	Identity X
Identity X	Identity N
Identity Z	Identity M
Identity N	Identity Z

Figure 3. Steps for the Markov chain rank fusion method (adopted from Sculley, 2006)

Step 1: Map the set of ranked lists to a single Markov chain, where one node of the chain represents single identity in the initial ranking lists.

Step 2: Compute the stationary distribution on the Markov chain.

Step 3: Rank the identities based on the stationary distribution. The node with the highest score in the stationary distribution is given the top rank, and so on until the node with the lowest score in the stationary distribution which is given the last rank.

sions of those heuristics. As an example, authors state that the considered Borda's method is based on the idea "more wins is better." They suggest to extend this notion and say "more wins against good players is even better" and by doing this to iteratively refine the ordering produced by a heuristic (Dwork, Kumar, Naor, & Sivakumar, 2001). Some Markov chain models for biometric rank aggregation can be viewed as the natural extensions of Borda's method, sorting by geometric mean or Copeland's method (Copeland, 1951) which will be defined below.

There are four specific Markov Chains (MCs): MC_1, MC_2, MC_3 and MC_4 proposed in (Dwork, Kumar, Naor, & Sivakumar, 2001). They differ in the way of how a ranking or a next state chosen. As was shown in (Monwar & Gavrilova, 2011), these Markov chains can be used for biometric rank aggregation. They are defined as follows in the context of biometric multi-modal system

(Monwar & Gavrilova, 2011). Suppose the current state of a Markov chain is *a*.

MC₁: Choose an identity *b* uniformly from multi-set of all identities that were ranked at least as high as *a* by some classifier. Probability to stay at *a*: ~ average rank of *a*.

MC₂: Choose a classifier *i* uniformly at random and pick uniformly at random from among the identities that the *i*-th classifier ranked at least as high as *a*.

MC₃: Choose a classifier *i* uniformly at random and pick uniformly at random an identity *b*. If *i*-th classifier ranked *b* higher than *a*, go to *b*. Otherwise, stay in *a*.

MC₄: Choose an identity *b* uniformly at random. If most classifiers ranked *b* higher than *a*, go to *b*. Otherwise, stay in *a*. So, the rank of *a* ~ # of "pairwise contests" *a* wins.

Among these four methods, only the last method satisfies the Condorcet criterion.

- **Copeland Method:** Sort the identities by the number of pairwise majority wins minus pairwise majority losses. Copeland's method satisfies the extended Condorcet condition, and is generalized by *MC₄* (Dwork, Kumar, Naor, & Sivakumar, 2001).

Figure 4 shows a Markov chain with its transition matrix build on *MC₄*. For this example, let us assume that four persons are to be classified by three classifiers/matchers.

But each classifier outputs only the first three results of their ranking list (i.e., each classifier outputs a partial list). From these partial lists, a full list has been created. The missing items in the list can be inserted randomly or by examining the partial lists. In this example, in the first list among the four subjects, only one is missing.

A subject (person) at the end of the list can be easily outputted without other consideration. As

the list of the first matcher already contains subject *a*, subject *b* and subject *c*, so the fourth subject is obviously subject *d*. Similarly, the already enlisted subjects in the list of second matcher are subject *b*, subject c and subject *d*. Hence the fourth entry in this list is subject *a*. According to the same method, the fourth entry in the list of the third matcher is subject *c* as subject *a*, subject *b* and subject *d* are already in the list.

In the case of more than one unlisted entries (subjects), two methods can be applied. The first method is the random method in which the subjects which are not listed in the partial list obtained from a matcher are positioned in the list by a random algorithm (Figure 5). The second method uses the relative positions of the unlisted subjects in the partial lists to place those (unlisted subjects) in the full ranking list. If the relative positions are not available, then a random algorithm is used (similar to the first method) to place the subjects in the final list (Figure 6).

Based on these full lists, a transition matrix is created. As there are four subjects considered in the example, so the transition matrix has four rows and four columns. The first row belongs to subject

Figure 4. Markov chain and the transition matrix constructed from three ranking lists based on MC₄

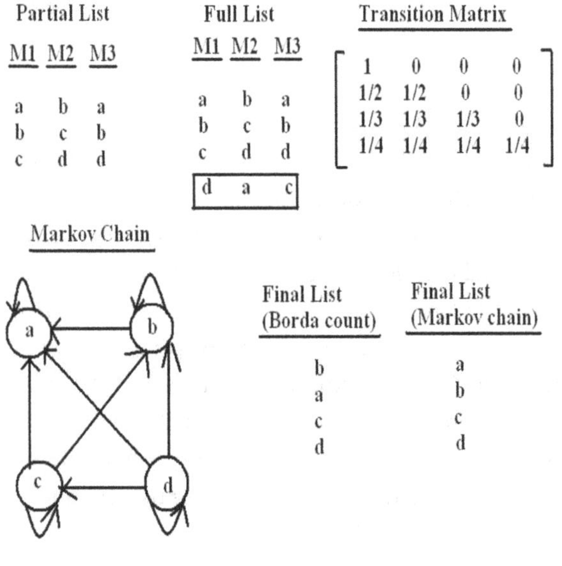

a, and similarly the second row, the third row and the fourth row belong to subject *b*, subject *c* and subject *d* respectively. In the same way, the first, second, third and fourth columns belong to subject *a*, subject *b*, subject *c* and subject *d* respectively. An entry '1' in the (1,1) position mean the only possible state to transition from state *a* is *a*. An entry '1/2' in position (2,1) means there is 50% probability to transition to state *a* from state *b*. Similarly, there is 50% probability to transition from state *b* from state *b*. In other words, from state *b*, only transition to state *a* and state *b* (itself) is possible. Further, from the fourth row of the transition matrix, it is clear that from the state *d*, transition to all other states is possible.

A Markov chain is constructed according to MC_4 from the transition matrix. Transition from one state to another state is shown using normal arrow. The final ranking list (which satisfies the Condorcet criterion) can be obtained by applying the Copeland method, i.e., by sorting the nodes in the majority graph (Markov chain) by out-degree minus in-degree. The figure also shows that if one applies the Borda count method to the lists,

he can obtain a final list, which does not satisfy the Condorcet criteria. This may also be the case for highest rank fusion as there is a tie between identity *a* and identity *b*. If this tie is broken randomly, there is 50% chances to select identity *b* as the winner, which is the violation of Condorcet criteria. Experimental results presented in the next subsections confirm that Markov chain rank fusion method is better than the other rank fusion methods, such as highest rank, Borda count, and logistic regression method.

Hence, this method can be a good solution to person identification problem for security critical multimodal biometric system, especially where the match score or feature sets are not available and the single biometric matchers can only output the ranking list of identities.

5. SAMPLE RESULTS OBTAINED THROUGH EXPERIMENTATIONS

In order to evaluate the performance of Markov chain based rank level fusion, experiments have

Figure 5. Solving partial list problem in Markov chain rank fusion method: random approach

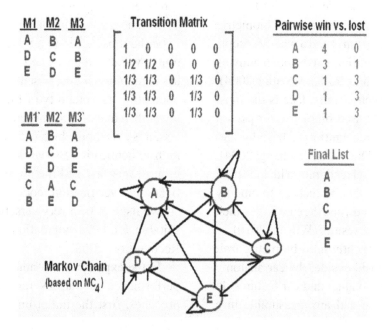

Figure 6. Solving partial list problem in Markov chain rank fusion method: relative positional approach

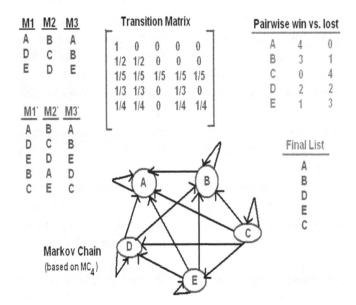

been conducted. In the experiments, face, ear, and iris are used as the unimodal biometrics. Researchers investigated different biometric identifiers based on several factors including application scenario, associated cost and availability of the identifiers (Ross, Nandakumar, & Jain, 2006). The choice is highly personal and depends on the individual system requirements, resource availability, training schedule, and other factors.

Face appearance is a highly popular biometric because it has been a natural and a widely acceptable way for recognizing humans by other humans (Bolle, Connell, Pankanti, Ratha, & Senior, 2004). Among all the biometric traits, face is the most common and heavily used biometric for person identification. Face recognition is friendly and non-invasive (Feng, Dong, Hu, & Zhang, 2004). The advantages of facial recognition include high public acceptance of this biometric, commonly available sensing devices, and the ease with which humans can verify the results (Wilson, 2010).

Ear is not as frequently used biometric trait as a face. On the positive side, the ear anatomy is unique to each individual and ear features do not change over time and are measurable in a

formalized way (Iannarelli, 1989). Given that the biometric is practical, identification by ear biometrics is promising because it is non-invasive, easy to obtain and can be represented in an image form (Burge & Burger, 1998). Further, ear images can be acquired in a similar manner to face images (i.e., the camera which is used for acquiring face can also be used for acquiring ear images) and can be efficiently used in surveillance scenarios.

Iris pattern recognition is generally considered to be the most accurate among all the biometric traits available today. The only disadvantage of iris recognition is cost of sensing equipment, and population acceptability of this biometric being lower than that of face or even ear one. Nevertheless, it is a method which in addition to providing high authentication also works for diverse sample groups, very fast, and flexibility for use in identification or verification modes, and thus has also shown itself to be a very versatile biometric also suitable for large population applications (Iris Recognition, 2003).

In the experiments conducted to validate the performance of various rank level fusion approaches, first the initial unimodal matching is done. The system then outputs the top-n matches of individual biometrics. Next, after selecting the appropriate rank fusion approach, the system outputs the final identification result.

5.1. Experimental Data

There are generally three types of experimental data available to biometric researcher today:

- True multi-modal database
- Virtual multi-modal database
- Synthetic multi-modal database

The true multi-modal database provides the best opportunity for a researcher to validate the methodology and is often a requirement for commercial system testing. In this database, each user provides all biometric samples for each individual biometric modality and they are stored as a raw data or templates. However, due to a number of factors, such as cost of database collection, privacy concerns in case the security of data would be compromised, and high limitations on size of the true multi-modal databases, a virtual multi-modal database is often used.

A virtual database allows the researcher to conduct experiments on a real data in very close to reality settings, without sacrificing the cost and time required to complete the project. The creation of virtual database is explained in Ross, Nandakumar, and Jain (2006), where it is comprised of the records created by pairing a user data from one unimodal database (e.g., face) with a user from another database (e.g., iris) (Ross, Nandakumar, & Jain, 2006). The creation of virtual users is normally relies on the assumption that different biometric traits of the same person are independent (Gavrilova & Monwar, 2009). An alternative is to consider true multimodal database with all biometrics coming from the same user. This is a bit more costly to obtain such database, and privacy issues become more prominent in accumulating

such data. This is one of the reasons why virtual multibiometric databases have been very popular.

A final type of biometric database is synthetic database. The trend has been on the rise due to high risks associated with storing both true and virtual databases on a single server or even in a distributed environment. The synthesis of biometric data was popularized in a book on Synthesis and Analysis in Biometric by Yanushkevich, Gavrilova, Wang, and Srihari (2007), which became World Scientific bestseller in 2007. In that comprehensive source, the strong motivation for the need of virtual biometric data is provided, with a variety of algorithms for virtual fingerprints, signatures, facial expressions and iris generation provided.

In experiments presented in Monwar and Gavrilova (2011), a virtual database approach has been used. The data from three different unimodal databases for iris, ear and face has been combined in a virtual database by random matching. For iris, the CASIA Iris Image Database (ver 1.0) maintained by the Chinese Academy of Science (Sino Biometrics, 2004) has been used. Iris images of this version of CASIA database were captured with a homemade iris camera. This iris database includes 756 black and white iris images from 108 eyes (hence 108 classes). For each eye, 7 images were captured in two sessions, where three samples are collected in the first session and four in the second session. The pupil regions of all iris images in CASIA-IrisV1 were automatically detected and replaced with a circular region of constant intensity to mask out the specular reflections (Sino Biometrics, 2004).

The ear images were obtained from the USTB database (USTB, 2012). This database contains ear images with illumination and orientation variation and individuals were invited to be seated 2m from the camera and change his/her face orientation. The images are 300 x 400 pixels in size. Due to the different orientation and image pattern, the ear images of this database needs normalization (USTB, 2012).

For face, the Facial Recognition Technology (FERET) database (Phillips, Moon, & Rauss, 1998) is used. It contains 24 facial image categories. FERET facial database was collected at George Mason University and the US Army Research Laboratory facilities and was recorded in 15 sessions between 1993 and 1996. All face images were recorded with a 35 mm camera and at last converted to 8-bit grey scale images. There are 14,051 images of 1199 person that are 256 x 384 in size. The face images in FERET database vary in subject pose, expression, and illumination (Phillips, Moon, & Rauss, 1998).

To build the virtual multimodal database for the multi-modal system, the following approach can be used. All the classes (subjects) of each datasets have been numbered. Then random selection has been made for the same classes from each three datasets. The images within the same class of three datasets are then paired to form a single class of our virtual multimodal database. Half of the classes are chosen for training purpose and the rest are used for testing purpose. Figure 7

shows a small portion of the virtual multimodal database created from CASIA iris dataset, FERET face dataset and USTB ear dataset.

To fully test the proposed multimodal biometric system performance, a second virtual multimodal database has been created. A sample of this database is shown in Figure 8. In this database, a public domain ear database (Perpinan, 1995) which contains 102 gray scale images (6 images for 17 subjects) has been used. The images were captured with a grey scale CCD camera Kappa CF 4 (focal 16 mm, objective 25.5 mm, f-number 1.4-16) using the program Vitec Multimedia Imager for VIDEO NT v1.52. Each raw image has a resolution of 384 x 288 pixels and 256-bit grey scales. The camera was at around 1.5 m from the subject. Six views of the left profile from each subject were taken under uniform lighting. Slight changes in the head position were allowed. There were 17 different subjects from the Faculty of Informatics of the Technical University of Madrid. The raw images were then cropped and rotated for uniformity (to a ratio height:width of 1.6), and

Figure 7. A small portion of the virtual multimodal database (Sino Biometrics, 2004; USTB, 2012; Phillips, Moon, & Rauss, 1998)

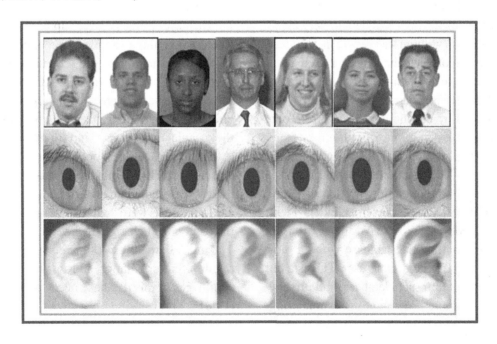

Figure 8. A portion of the second virtual multimodal database (Perpinan, 1995; University of Essex, 2008; Dobes, Martinek, Skoupil, Dobesova, & Pospisil, 2006)

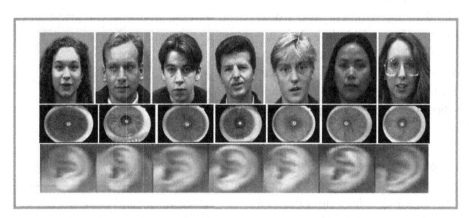

slightly brightened (gamma = 1.5 approx.), using the xv program in a Linux system (Perpinan, 1995).

The face data in the second virtual multimodal database is from the University of Essex, UK Computer Vision Science Research Project (University of Essex, 2008). There are 395 subjects in this face dataset with each having 20 face images and almost all of them are undergraduate students (age range is 18-20). Each image has a resolution of 180 x 200 pixels. The subjects are both male and female and the background of the image is plain green. The lighting and expression variations in this database between subjects is minimal (University of Essex, 2008).

The iris dataset in the second database is from the Department of Computer Science at Palacky University in Olomouc, Czech Republic (Dobes, Martinek, Skoupil, Dobesova, & Pospisil, 2006). This iris database contains 3 iris images of left eye and 3 iris images of right eye of 64 subjects.

The iris image size is 24 bits, RGB with resolution of 576 x 768 pixels. The irises were scanned by TOPCON TRC50IA optical device connected with SONY DXC-950P 3CCD camera (Dobes, Martinek, Skoupil, Dobesova, & Pospisil, 2006).

5.2. Experimental Results

This section summarizes and extends results presented in Monwar and Gavrilova (2011). The goal of this is to establish the superiority of the Markov chain based rank level fusion method with other methods. After experimentation, the results have been analyzed by plotting the recognition values on different biometric system performance curves. For the proposed rank level fusion, a Cumulative Match Characteristic (CMC) curve is used to summarize the identification rate at different rank values. As rank level fusion method can only be used for identification, the identification rate has been used which is the proportion of times the identity determined by the system is the true identity of the user providing the query biometric sample. If the biometric system outputs the identities of the top x candidates, the rank-x identification rate is the proportion of times the true identity of the user is contained in the top-n candidate identities.

Figure 9(a-c) shows CMC curves for the sample three unimodal matchers utilizing the first virtual multimodal database. Among the three unimodal matchers, iris matchers produce the best results with the 93.21% rank-1 identification rate. Rank-1 identification rates for face and ear are 92.03% and 87.16% respectively.

Figure 10 shows the CMC curves for four rank level fusion approaches applied on the first virtual multimodal database. Highest rank, Borda count, logistic regression and Markov chain approaches to rank level fusion have been applied in this experiment and the best rank-1 identification rates through Markov chain approach (97.96%) has been obtained. Among the other three, logistic regression approach is the best (almost 95.93%). The results can be explained as follows. As the performances of our individual matchers are not equal, hence the highest rank and Borda count approaches have not produced satisfactory classification results. Borda count rank fusion approach produced 94.81% rank-1 identification rate whereas, the highest rank fusion approach produced 93.89% rank-1 identification rate.

To properly evaluate the face, ear and iris based multimodal biometric system, performance of the system has been tested the system on the second virtual multimodal database. Figure 11 shows the CMC curves for face, ear, and iris matchers. Among the three unimodal matchers, for the second virtual multimodal database, face matcher produced the best result with a 91.84% rank-1 identification rate. For iris and ear, the rank-1 identification rate is 87.13% and 81.67% respectively. These results differ from the results obtained from the first virtual multimodal database as the three individual datasets differ a lot in quality. In the second virtual multimodal database, the face images are very clear with very limited illumination and pose changes. On the other hand, the quality of the ear dataset is not good and the inter-class variations among the ear images are very limited. Thus, this ear dataset produced lower rank-1 identification rate. Similarly, the identification rates of the iris images are lower. These factors have influenced the outcomes of the unimodal matchers.

Figure 9. CMC curves for unimodal biometrics: (a) for face, (b) for ear, and (c) for iris

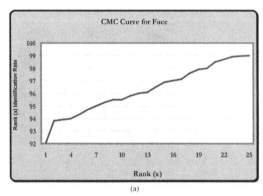

(a)

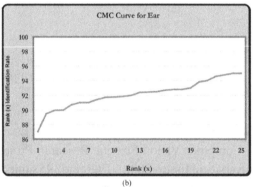

(b)

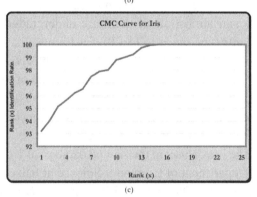

(c)

Figure 12 shows the CMC curves for four rank level fusion approaches along with the best unimodal matcher (face) applied to the second virtual multimodal database. Similar to the previous experimentation, highest rank, Borda count, logistic regression, and Markov chain approaches for rank fusion have been applied. Among all of

Figure 10. CMC curves for four rank fusion approaches applied on the virtual multimodal database

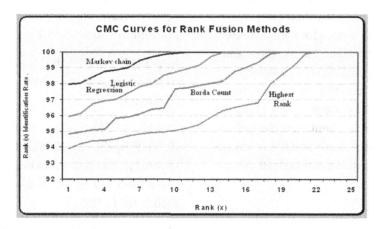

Figure 11. CMC curves for unimodal matchers applied on the second virtual multimodal database

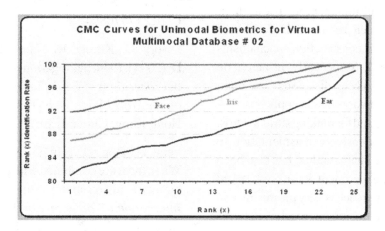

Figure 12. CMC curves for four rank fusion approaches and face unimodal matcher applied on the second virtual multimodal database

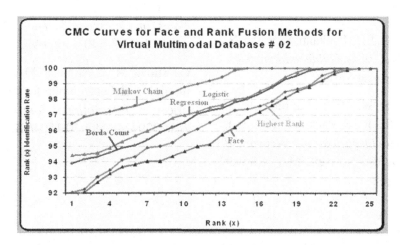

these, Markov chain approach outperforms other with a 96.45% rank-1 identification rate.

Rank-1 identification rates for logistic regression, Borda count and highest rank approaches are 94.41%, 93.89%, and 92.03%, respectively. As the quality of the ear and iris images in the respective datasets are comparatively poor, the highest rank and Borda count rank fusion approaches have produced non-satisfactory results compared to the first virtual multimodal database.

6. SUMMARY

In this chapter, Markov chain based rank level fusion method for multimodal biometric authentication system has been discussed. Definition of Markov chain and its construction mechanisms have been presented. Some early research on Markov chain has also been discussed. The main concentration was on two particular applications—for Web page ranking and for ranking similar items. Description of an extensive experimentation to evaluate the Markov chain method has also been presented. From the experiment, it can be observed that Markov chain approach is very promising and can outperform other fusion approaches in terms of biometric rank aggregation on virtual database.

REFERENCES

Bolle, R. M., Connell, J. H., Pankanti, S., Ratha, N. K., & Senior, A. W. (2004). *Guide to biometrics.* New York, NY: Springer-Verlag.

Burge, M., & Burger, W. (1998). Ear biometrics. In Jain, A. K., Bolle, R., & Pankanti, S. (Eds.), *Biometrics: Personal Identification in Networked Society* (pp. 273–286). Norwell, MA: Kluwer Academic Publishers.

Condorcet, M.-J. (1785). *Essai sur l'application de l'analyse a la probabilite des decisions rendues a la pluralite des voix.* Paris, France: Academic Press.

Copeland, A. H. (1951). *A reasonable social welfare function.* Ann Arbor, MI: University of Michigan.

Dobeš, M., Martinek, J., Skoupil, D., Dobešová, Z., & Pospíšil, J. (2006). Human eye localization using the modified Hough transform. *Optik (Stuttgart), 117*(10), 468–473. doi:10.1016/j.ijleo.2005.11.008.

Dutkowski, J., & Gambin, A. (2007). On consensus biomarker selection. *BioMed Central Bioinformatics, 24*(8), S5. doi:10.1186/1471-2105-8-S5-S5 PMID:17570864.

Dwork, C., Kumar, R., Naor, M., & Sivakumar, D. (2001). Rank aggregation methods for the web. In *Proceedings of Tenth International Conference on the World Wide Web (WWW)*, (pp. 613-622). Hong Kong, China: IEEE.

Feng, G., Dong, K., Hu, D., & Zhang, D. (2004). When faces are combined with palmprint: A novel biometric fusion strategy. In *Proceedings of First International Conference on Biometric Authentication*, (pp. 701-707). Hong Kong, China: IEEE.

Gambin, A., & Pokarowski, P. (2001). A combinatorial aggregation algorithm for stationary distribution of a large Markov chain. *Lecture Notes in Computer Science, 2138*, 384–387. doi:10.1007/3-540-44669-9_38.

Gavrilova, M., & Monwar, M. M. (2009). Fusing multiple matchers' outputs for secure human identification. *International Journal of Biometrics, 1*(3), 329–348. doi:10.1504/IJBM.2009.024277.

Grinstead, C. M., & Snell, J. L. (1997). *Introduction to probability* (2nd ed.). Providence, RI: American Mathematical Society.

Iannarelli, A. (1989). *Ear identification*. Fremont, CA: Paramont Publishing Company.

Iris Recognition. (2003). *Iris technology division*. Cranbury, NJ: LG Electronics USA.

Kemeny, J. G., Snell, J. L., & Thompson, G. L. (1974). *Introduction to finite mathematics* (3rd ed.). Englewood Cliffs, NJ: Prentice-Hall.

Markov, A. A. (1906). Extension of the limit theorems of probability theory to a sum of variables connected in a chain. In R. Howard (Ed.), Dynamic Probabilistic Systems, Volume 1: Markov Chains. Hoboken, NJ: John Wiley and Sons.

Monwar, M. M., & Gavrilova, M. (2011). Markov chain model for multimodal biometric rank fusion. In *Proceedings of Signal, Image and Video Processing*. Springer. doi:10.1007/s11760-011-0226-8.

Perpinan, C. (1995). *Compression neural networks for feature extraction: Application to human recognition from ear images*. (M.Sc. Thesis). Technical University of Madrid. Madrid, Spain.

Phillips, P. J., Moon, H., & Rauss, P. (1998). The FERET database and evaluation procedure for face recognition algorithms. *Image and Vision Computing, 16*(5), 295–306. doi:10.1016/S0262-8856(97)00070-X.

Ross, A., Nandakumar, K., & Jain, A. K. (2006). *Handbook of multibiometrics*. New York, NY: Springer-Verlag.

Sculley, D. (2006). Rank aggregation for similar items. Report. New York, NY: Data Mining and Research group of Yahoo.

Sino Biometrics. (2004). *CASIA: Casia iris image database*. Retrieved from www.sinobiometrics.com

University of Essex. (2008). *Face database*. Retrieved from http://cswww.essex.ac.uk/mv/allfaces/index.html

USTB. (2012). *Ear database, China*. Retrieved from http://www.ustb.edu.cn/resb/

Wilson, C. (2010). *Vein pattern recognition: A privacy-enhancing biometric*. Boca Raton, FL: CRC Press. doi:10.1201/9781439821381.

Yanushkevich, S., Gavrilova, M., Wang, P., & Srihari, S. (2007). *Image Pattern Recognition: Synthesis and Analysis in Biometrics*. New York, NY: World Scientific Publishers.

Chapter 7
Fuzzy Fusion for Multimodal Biometric

ABSTRACT

Fuzzy logic is a mathematical tool that can provide a simple way to derive a conclusion with the presence of noisy input information. It is a powerful intelligent tool and used heavily in many cognitive and decision-making systems. In this chapter, fuzzy logic-based fusion approach for multimodal biometric system is discussed. After discussing the basics of fuzzy logic, the fuzzy fusion mechanism in the context of a multimodal biometric system is illustrated. A brief discussion on the research conducted for fuzzy logic-based fusion in different application domains is also presented. The biggest advantage of the system is that instead of binary "Yes"/"No" decision, the probability of a match and confidence level can be obtained. A fuzzy fusion-based biometric system can be easily adjusted by controlling weight assignment and fuzzy rules to fit changing conditions. Some results of experimentations conducted in a recent research investigation on two virtual multimodal databases are presented. The discussion on the effect of incorporating soft biometric information with the fuzzy fusion method to make the system more accurate and robust is also included.

1. INTRODUCTION

In chapter six, the Markov chain based rank level fusion method has been introduced. The basics of Markov chain have been discussed and its construction mechanism in the context of multimodal biometric rank fusion has been shown. This method demonstrates a number of advantages over other rank fusion approaches in terms of recognition performance. Furthermore, this method satisfies the Condorcet criterion, which is essential in any fair rank information fusion process. In this chap-

DOI: 10.4018/978-1-4666-3646-0.ch007

ter, another new biometric fusion approach based on fuzzy logic is discussed and hence named as *fuzzy fusion* for multibiometrics.

Fuzzy fusion method is one of sub-branches of information fusion, which has recently emerged as information consolidation tool. Most fuzzy fusion methods reported in the literature are developed for areas such as automatic target recognition, biomedical image fusion and segmentation, gas turbine power plants fusion, weather forecasting, aerial image retrieval and classification, vehicle detection and classification, and path planning. In the context of biometric authentication, fuzzy logic based fusion approach has recently been used for quality based biometric information consolidation process. In Monwar, Gavrilova, and Wang (2011), the fuzzy fusion method is utilized in multimodal biometric system. The advantage of fuzzy fusion method is that it utilizes both match score and rank information from unimodal biometrics. Also, unlike with traditional systems returning only binary (Yes/No) decision, the level of confidence in recognition outcomes of the multimodal system can be obtained using this method.

2. FUZZY LOGIC BASICS

Fuzzy logic refers to the theories and technologies that employ fuzzy sets, which are classes with un-sharp boundaries (Pedrycz & Gomide, 1998). The idea of fuzzy sets was introduced in 1965 by Professor Lotfi A. Zadeh from the University of California, Berkeley (Zadeh, 1965). The core technique of fuzzy logic is based on following four basic concepts (Wang, 2009):

- **Fuzzy Sets:** A fuzzy set is a set with a smooth boundary. Fuzzy set theory generalizes the classical set theory to allow partial membership (Harb & Al-Smadi, 2006).
- **Linguistic Variable:** A linguistic variable in one which allows its value to be described both qualitatively by a linguistic

term and quantitatively by a corresponding membership function (which represents the meaning of the fuzzy set) (Harb & Al-Smadi, 2006).
- **Possibility Distributions:** Assigning a fuzzy set to a linguistic variable constrains the value of the variable: it generalizes the difference between possible and impossible to a degree called the possibility (Pedrycz & Gomide, 1998).
- **Fuzzy Rules:** Fuzzy rule (or the fuzzy if-then rule) is the most widely used technique developed using fuzzy sets and has been applied to many disciplines. Some of the applications of fuzzy rules include control (robotics, automation, tracking, consumer electronics), information systems (DBMS, information retrieval), pattern recognition (image processing, machine vision), decision support (adaptive HMI, sensor fusion), and cognitive informatics (Pedrycz & Gomide, 1998).

The development of fuzzy rule-based inference consists of three steps – fuzzification, inference and defuzzification (Figure 1) (Zadeh, 1965). In the *fuzzification* step, fuzzy variables and their membership functions are defined, i.e., the degree to which the input data match the condition of the fuzzy rules have been calculated. In the *inference* step, fuzzy rules have been developed and those rules outcome based on their matching degree has been calculated. In the *defuzzification* step, the fuzzy conclusion is converted into a discrete one (Zadeh, 1965).

3. RESEARCH ON FUZZY LOGIC-BASED FUSION

Fuzzy logic is indeed one of the fascinating areas on the edge between cognitive science and decision making. Utilizing principles of fuzzy logic for information fusion allows to emulate

Figure 1. Fuzzy rule-based inference system (adopted from Zadeh, 1965)

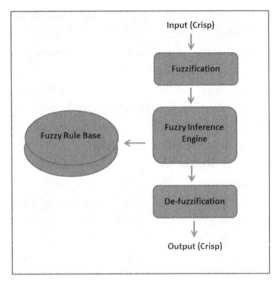

the abstract reasoning and complex human intelligence processes by the means of a range of values between yes or no, or true and false. In a machine language, it is a difference between 0 or 1, and in biometric terms—between accepted and refused identity, or granted or denied access to a secure premises or a facility.

One of the first works in this domain which has significance for decision-making is 1999 work by Solaiman et al. (Solaiman, Pierce, & Ulaby, 1999). The authors proposed a fuzzy-based multisensor data fusion classifier for to be used in a geo-spatial and remote sensing domain for land cover classification. Their classifier provided a tool for integration of multisensor and contextual information. The authors introduced the Fuzzy Membership Maps (FMMs) to represent different thematic classes based on a priori information obtained from sensors. The FMMs were next iteratively updated using spatial contextual information. The fuzzy logic allowed their proposed classifier to integrate multisensor and a priori information.

In another study published in 2001, Kim et al. (Kim, Kim, Lee, & Cho, 2001) developed a new method for vehicle classification using fuzzy logic. In their algorithm, the vehicle weight and speed were used as the inputs to the fuzzy logic module. The output of the fuzzy logic module was a weighting factor to modify the vehicle length calculated using the raw sensor outputs. The modified length was the input to the vehicle classification block, and the final classification result was generated. With their experimental results, they showed that their classification algorithm using the fuzzy logic significantly reduces the errors in vehicle classification.

One of the key contributions of the fuzzy fusion method is the work of Wang et. al. in 2007 (Wang, Dang, Li, & Li, 2007), where authors used fuzzy fusion for multimodal medical image fusion. To overcome the problem of blurriness of the most medical images, the authors proposed a new method of medical image fusion using fuzzy radial basis function neural networks (Fuzzy-RBFNN), which is functionally equivalent to T-S fuzzy model. Genetic algorithm was used to train the networks which performed a global exploration of the search space and used several heuristics to avoid the networks getting trapped in local optima. Experiment results, conducted on 20 groups CT (Computed Tomography) and MRI (Magnetic Resonance Imaging) images of the head, showed that the proposed approach outperformed the gradient pyramid based image fusion in both visual effect and objective evaluation criteria, especially for blurry images.

The follow-up work by Wang (2009) provided a fundamental theoretical backing to a formal knowledge system theory and its cognitive informatics foundations. In additional to a strong theoretical foundation for cognitive system development, it also provided knowledge representation tool capable of representing concepts in multiple ways as well as to visualize the dynamic concept networks by the means of machine learning based on concept algebra principles.

In 2010, Deng et al. (Deng, Su, Wang, & Li, 2010), proposed a data fusion method based on fuzzy set theory and Dempster-Shafer evidence theory (Shafer, 1976) for automatic target recognition, which could deal with uncertain data in a flexible manner. The authors represented both the individual attribute of target in the model database and the sensor observation as fuzzy membership function and constructed a likelihood function to represent fuzzy data collected from sensors. Sensor data from different sources was used based on the Dempster combination rule (Chang, Bowyer, & Barnabas, 2003).

Very recently, in another research on applying fuzzy fusion in the biomedical imaging research domain, Chaabane and Abdelouahab (2011) proposed a system of fuzzy information fusion framework for the automatic segmentation of human brain tissues using T2-weighted (T2) and Proton Density (PD) images. Their system consisted of the computation of fuzzy tissue maps in both images using Fuzzy C-means algorithm. The effectiveness of their system was established experimentally. The results from their experiments showed the applicability of the data fusion in the medical imaging field.

From the examples above, we can observe the strong trend on utilizing intelligence decision making approaches, such as cognitive intelligence and fuzzy logic in a field with high complexity and variability of data types and constraints. The areas are defined by a large number of attributes, and decisions are highly sensitive to the parameters in such a high-dimensional problem space. It has emerged as a trend over last decade that areas involving geospatial, medical, oil and gas, and biometric data would benefit most from using described intelligent decision-making methodology.

4. FUZZY FUSION OF BIOMETRIC INFORMATION

Now, let us take a closer look at the way of how fuzzy logic can be utilized in biometric security domain, including both conceptual and practical aspects of such integration.

Figure 2 shows a data flow chart for a sample fuzzy fusion module, which is a fuzzy rule-based inference system. Similar to the experiments conducted for the evaluation of Markov chain based rank fusion method (Monwar & Gavrilova, 2011); this fuzzy fusion method also utilizes face, ear and iris biometric information. At first, the three matchers compare the three input biometric data with the stored templates and produce ranking based on the similarity/distance scores. Markov chain based rank fusion approach only utilizes rank information of a multimodal biometric system, on the other hand fuzzy fusion based biometric rank fusion uses rank as well as match score for biometric information consolidation.

The initial input to this fuzzy fusion module is the individual similarity scores and the average similarity score for a person. The output of this module is the identification decision of the multimodal biometric system.

The fuzzy inference mechanism is the centre of the fuzzy fusion module. As discussed in the previous section, the first step for fuzzy inference is fuzzification where the input is modelled as fuzzy variables.

Next, fuzzy-rule inference is defined based on degree of which fuzzy variables match the rules.

Finally, a discrete decision of the type "Yes" or "No", "Accept" and "Deny", or "genuine" or "Impostor" is given by the defuzzification module. Since the outcome of fuzzy decision-making module can have a range between 0 and 1, this decision can be transformed into *degree of confidence*, or probability that the user is a genuine user or impostor.

Figure 2. Fuzzy fusion module flowchart for a multimodal biometric system

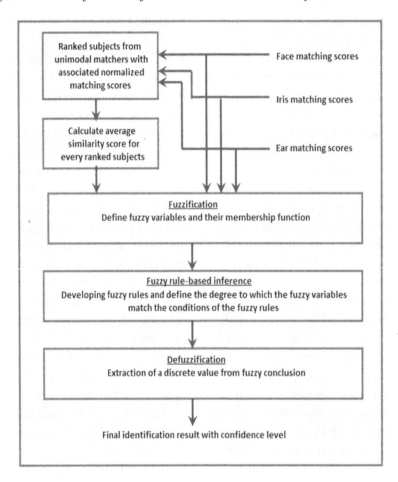

Score normalization is necessary for applying the fuzzy fusion technology. We used min-max normalization technique for score normalization to obtain all match score values in the range of 0 to 1.

For a multimodal system, suppose, s_j^i denote the *i*-th match score output by the *j*-th matcher, i $=1, 2, ..., N$, where N is the number of subjects enrolled in the system, and $j = 1, 2, 3$. Min-max normalization preserves the original distribution of scores and transforms all the scores into a normalized range [0, 1]. The min-max normalized score ns_j^t for the test score s_j^t is often defined as follows (Ross, Nandakumar, & Jain, 2006):

$$ns_j^t = \frac{s_j^t - \min_{i=1}^{n} s_j^i}{\max_{i=1}^{n} s_j^i - \min_{i=1}^{n} s_j^i} \qquad (7.1)$$

Assume there are N subjects enrolled in the database and among them, K users appeared in the ranking lists of three matches, i.e. $K \leq N$ AND $n \leq K \leq 3n$ as only top-*n* ranked subjects produced by each matcher. Let the number of biometric traits and matchers be M, i.e. M = 3. Let $s_{k,m}$ be the match score generated for subject k by m matcher, $s_{k,m} \geq 0$ and $s_{k,m} \leq 1$ after normalization. Thus, the average similarity score for a particular subject s_k can be obtained by the following equation:

$$s_k = \frac{1}{M} \sum_{m=1}^{M} s_{k,m} \qquad (7.2)$$

There are only top-n ranked matches produced by each matcher, but average match scores for more than n subjects may be available, since some identifiers may not be included in all of the rank lists (maximum $3n$; i.e. when all three matchers output ranking list based on the match score where no identifiers). In this case, a provision for the fusion module is employed to collect the absent matching scores from the matching modules. For this purpose, initially the fusion module compares the identity presents in all three ranking lists and brings necessary matching scores from the matching module for comparison. Later the resultant average similarity score and the three match scores of three classifiers are used as inputs to the proposed fuzzy inference system.

After obtaining these fuzzy variables, the fuzzy membership function is defined, which is the degree to which the input data match the condition of the fuzzy rules. Suppose the multimodal database used is a virtual multimodal database based on three different datasets collected from three different sources, so .95 (out of maximum 1) similarity score is considered as very good matching score. Thus, the fuzzy linguistic variables, high (H), medium (M), and low (L) are defined as follows (Monwar, Gavrilova, & Wang, 2011):

H, when $s \geq 0.95$

M, when $0.80 \leq s \leq 0.94$ $\qquad (7.3)$

L, when $s \leq 0.79$

Once the fuzzy variables are adequately mapped into the membership functions, the task is to develop the fuzzy rules, which are elaborated on the basis of individual biometric matching performances and on the robustness of the biometric traits. This step is necessary to obtain the confidence of the final recognition outcome from the system, which is one of my motivations to utilize this fusion method. For the fuzzy inference system, 51 rules are developed which are explained in Figure 3 (Monwar, Gavrilova, & Wang, 2011).

For these rules, the average match score as well as the performances of individual matchers is considered. The reason to use the individual match scores is that the virtual biometric databases are used in this system. They are obtained from different sources and hence are of different quality. Further, different matching algorithms are used for this system: the fisherimage technique for face and ear biometrics, and Hough transform with Gabor wavelet and Hamming distance techniques for iris biometric. Thus, different results from three different matchers are obtained which allow to put different confidences on matching results. Among the four inputs to the fuzzy inference system (average match score and three individual match scores), the highest confidence on the average match score in the fuzzy rules is assigned.

Based on the previously evaluated biometric performance (Monwar, Gavrilova, & Wang, 2011; Monwar & Gavrilova, 2011, 2009), among the three individual match scores, we put the highest confidence on the score obtained from the iris matcher and the lowest confidence on the score obtained from the ear matcher.

With the four parameters, i.e. the average score and the three unimodal matcher's scores, there are 81 alternatives. Among these 81 alternatives, only the 51 possibilities are possible, as according to the definition of the fuzzy linguistic variables, the alternatives such as—

If AS = 'L', FS = 'M', IS = 'H' and ES = 'M' and

If AS = 'M', FS = 'H', IS = 'H' and ES = 'H'

(AS = Average score; FS = Face matcher's score; IS = iris matcher's score; ES = Ear matcher's score)

Figure 3. Fuzzy rules for the proposed fuzzy fusion method (Monwar, Gavrilova, & Wang, 2011)

1. If AS = 'H', FS = 'H', IS = 'H' and ES = 'H', then 'SI'	26. If AS = 'M', FS = 'L', IS = 'H' and ES = 'M', then 'WI'
2. If AS = 'H', FS = 'H', IS = 'H' and ES = 'M', then 'SI'	27. If AS = 'M', FS = 'L', IS = 'H' and ES = 'L', then 'WI'
3. If AS = 'H', FS = 'H', IS = 'M' and ES = 'H', then 'SI'	28. If AS = 'M', FS = 'L', IS = 'M' and ES = 'H', then 'WI'
4. If AS = 'H', FS = 'H', IS = 'M' and ES = 'M', then 'SI'	29. If AS = 'M', FS = 'L', IS = 'M' and ES = 'M', then 'WI'
5. If AS = 'H', FS = 'M', IS = 'H' and ES = 'H', then 'SI'	30. If AS = 'M', FS = 'L', IS = 'M' and ES = 'L', then 'WI'
6. If AS = 'H', FS = 'M', IS = 'H' and ES = 'M', then 'SI'	31. If AS = 'M', FS = 'L', IS = 'L' and ES = 'H', then 'NI'
7. If AS = 'H', FS = 'M', IS = 'M' and ES = 'H', then 'WI'	32. If AS = 'M', FS = 'L', IS = 'L' and ES = 'M', then 'NI'
8. If AS = 'M', FS = 'H', IS = 'H' and ES = 'M', then 'WI'	33. If AS = 'L', FS = 'H', IS = 'H' and ES = 'L', then 'WI'
9. If AS = 'M', FS = 'H', IS = 'H' and ES = 'L', then 'WI'	34. If AS = 'L', FS = 'H', IS = 'M' and ES = 'L', then 'NI'
10. If AS = 'M', FS = 'H', IS = 'M' and ES = 'H', then 'WI'	35. If AS = 'L', FS = 'H', IS = 'L' and ES = 'H', then 'NI'
11. If AS = 'M', FS = 'H', IS = 'M' and ES = 'M', then 'WI'	36. If AS = 'L', FS = 'H', IS = 'L' and ES = 'M', then 'NI'
12. If AS = 'M', FS = 'H', IS = 'M' and ES = 'L', then 'WI'	37. If AS = 'L', FS = 'H', IS = 'L' and ES = 'L', then 'NI'
13. If AS = 'M', FS = 'H', IS = 'L' and ES = 'H', then 'WI'	38. If AS = 'L', FS = 'M', IS = 'H' and ES = 'L', then 'WI'
14. If AS = 'M', FS = 'H', IS = 'L' and ES = 'M', then 'WI'	39. If AS = 'L', FS = 'M', IS = 'M' and ES = 'L', then 'NI'
15. If AS = 'M', FS = 'H', IS = 'L' and ES = 'L', then 'WI'	40. If AS = 'L', FS = 'M', IS = 'L' and ES = 'H', then 'NI'
16. If AS = 'M', FS = 'M', IS = 'H' and ES = 'H', then 'WI'	41. If AS = 'L', FS = 'M', IS = 'L' and ES = 'M', then 'NI'
17. If AS = 'M', FS = 'M', IS = 'H' and ES = 'M', then 'WI'	42. If AS = 'L', FS = 'M', IS = 'L' and ES = 'L', then 'NI'
18. If AS = 'M', FS = 'M', IS = 'H' and ES = 'L', then 'WI'	43. If AS = 'L', FS = 'L', IS = 'H' and ES = 'H', then 'WI'
19. If AS = 'M', FS = 'M', IS = 'M' and ES = 'H', then 'WI'	44. If AS = 'L', FS = 'L', IS = 'H' and ES = 'M', then 'NI'
20. If AS = 'M', FS = 'M', IS = 'M' and ES = 'M', then 'WI'	45. If AS = 'L', FS = 'L', IS = 'H' and ES = 'L', then 'NI'
21. If AS = 'M', FS = 'M', IS = 'M' and ES = 'L', then 'WI'	46. If AS = 'L', FS = 'L', IS = 'M' and ES = 'H', then 'NI'
22. If AS = 'M', FS = 'M', IS = 'L' and ES = 'H', then 'WI'	47. If AS = 'L', FS = 'L', IS = 'M' and ES = 'M', then 'NI'
23. If AS = 'M', FS = 'M', IS = 'L' and ES = 'M', then 'NI'	48. If AS = 'L', FS = 'L', IS = 'M' and ES = 'L', then 'NI'
24. If AS = 'M', FS = 'M', IS = 'L' and ES = 'L', then 'NI'	49. If AS = 'L', FS = 'L', IS = 'L' and ES = 'H', then 'NI'
25. If AS = 'M', FS = 'L', IS = 'H' and ES = 'H', then 'WI'	50. If AS = 'L', FS = 'L', IS = 'L' and ES = 'M', then 'NI'
	51. If AS = 'L', FS = 'L', IS = 'L' and ES = 'L', then 'NI'

AS = Average score; FS = Face matcher's score; IS = iris matcher's score; ES = Ear matcher's score
SI = Strongly identified; WI = Weakly identified; NI = Not identified.

—are not possible.

At the final stage of this fuzzy inference system, a single scalar output suitable for the final classification by combining the results produced by all fuzzy rules is obtained. Figure 4 shows the steps for this fuzzy fusion method.

The system performance of fuzzy fusion utilizing *soft biometric* information is also tested. Soft biometric was briefly introduced in Chapter 2,

and includes such personal data as height, weight, race, gender, hair, and eye colour, etc. The notion of soft biometric has only recently made its way into biometric research, and a few systems to date tried to incorporate it or to estimate the benefits of using such information.

In the case of incorporating soft biometrics, the fuzzy inference engine is a two input one output system, unlike in the first case, where the

Figure 4. Steps for fuzzy fusion method

Step 1: Normalize all match scores to a value in between 0 to 1.

Step 2: Calculate average match scores.

Step 3: Define linguistic variables and their membership function.

Step 4: Create fuzzy rules that describe the relations between the variables.

Step 5: Establish a defuzzification process to get the final outcome as an identification decision with the level of confidence on that decision.

---

fuzzy inference engine is a four input one output system. The average similarity score is obtained by a different formula which is shown below:

$$s_k = \frac{1}{M} \sum_{m=1}^{M} w_m s_{k,m} \qquad (7.4)$$

where w_m is a weight for the *m*-th matcher and

$$\sum_{m=1}^{3} w_m = 1.0.$$

Weights on different matching scores are applied based on the same consideration, i. e., based on the expectations and evaluations of individual biometric matcher performance. The following weights for the three match scores are assigned:

$$\begin{cases} w_m = 0.45, \text{ for iris match score} \\ w_m = 0.30, \text{ for face match score} \\ w_m = 0.25, \text{ for ear match score} \end{cases} \qquad (7.5)$$

$$s_k = \frac{1}{M} \sum_{m=1}^{M} w_m s_{k,m}$$

The second input to the fuzzy inference system is the average soft biometric score. The same

procedure is applied to the latter as the average primary biometrics match score.

Three soft biometrics values—gender, ethnicity, and eye colour—were used in the system. Suppose, the number of soft biometric used in this system is S and $soft_{k,i}$ is the value of *k-th* user for *i-th* soft biometric, where $i > 0$ and $i \leq S$. One can use only the Boolean value for soft biometrics, i.e., either $soft_{k,i} = 0$ or $soft_{k,i} = 1$. One possible assignment of the weights for this soft biometrics is shown below:

$$\begin{cases} w_i = 0.50, \text{ for gender} \\ w_i = 0.30, \text{ for ethnicity} \\ w_i = 0.20, \text{ for eye color} \end{cases} \qquad (7.6)$$

The average soft biometric score for a particular subject $soft_{k,i}$ can then be obtained by the following equation:

$$soft_k = \sum_{i=1}^{S} w_i soft_{k,i} \qquad (7.7)$$

where w_i is the weight for the *i-th* soft biometric

$$\sum_{i=1}^{3} W_i = 1.0.$$

Once the average weighted match score and average weighted soft biometric score are obtained, they are used as input to the fuzzy inference engine. In this case, it is advisable to put less confidence on soft biometric score, as soft biometrics information are not fully reliable and can be altered easily by the impostor. For the two inputs one output fuzzy inference system, the rules shown in Figure 5 are considered. Experimental results indicate that the inclusion of the soft biometric information does not improve the recognition performance by a significant amount. Also, privacy problem arises when soft biometrics information is used. For this reason, using soft biometrics in multimodal biometric security systems is not the first choice in real world high security systems.

5. EXPERIMENT RESULTS FOR FUZZY FUSION OF BIOMETRIC INFORMATION

In the experiment, the same two datasets which were used in the experiments involving Markov chain-based rank fusion are used. Here, comparison has been made on fuzzy fusion approach with unimodal matchers, with the rank fusion

Figure 5. Fuzzy rules for the fuzzy fusion method utilizing soft biometric information

1. If AS = 'H' and SS = 'H', then 'SI'
2. If AS = 'H' and SS = 'M', then 'SI'
3. If AS = 'H' and SS = 'L', then 'WI'
4. If AS = 'M' and SS = 'H', then 'WI'
5. If AS = 'M' and SS = 'M', then 'WI'
6. If AS = 'M' and SS = 'L', then 'WI'
7. If AS = 'L' and SS = 'H', then 'WI'
8. If AS = 'L' and SS = 'M', then 'NI'
9. If AS = 'L' and SS = 'L', then 'NI'

AS = Average scores; SS = Soft biometrics score
SI = Strongly identified; WI = Weakly identified; NI = Not identified.

approaches and with Match score and decision fusion approaches which is shown in through Figures 6-11.

Figure 6 shows the ROC curves for the unimodal matchers and for the fuzzy fusion approach which are obtained through the experimentation with my first virtual multimodal database.

For a FAR of 0.1%, a 95.82% GAR (Genuine Accept Rate) was recorded, which is equivalent to (1 - FRR). For the unimodal matchers, for the same FAR, i.e., 0.1%, the GARs for face, ear and iris are 84.03%, 80.56% and 91.56% respectively.

Figure 6. ROC curves for unimodal biometrics and for fuzzy fusion

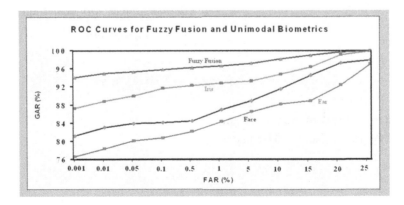

Figure 7. ROC curves for fuzzy fusion and different rank fusion approaches

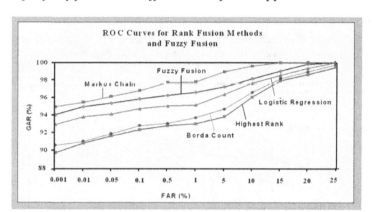

For the first virtual multimodal database, fuzzy fusion outperformed highest rank, Borda count and logistic regression methods (Figure 7). For a FAR of 0.1%, the GAR of highest rank fusion method is 92.31%, the GAR of Borda count method is 92.79%, and the GAR of logistic regression method is 94.71%. Among all of these fusion approaches, Markov chain method gave the best GAR of 96.75% for the same FAR. Although the recognition performance of the new fuzzy fusion method is not as good as Markov chain based rank fusion method, this method gives us the level of confidence on the recognition outcomes, which is important in some application areas. Also, the

fuzzy rules of this fusion method can be extended to make decisions on "Strictly Not Identified" subjects for some application areas, such as access to a very restricted area.

In order to efficiently evaluate the proposed system and to compare with other well-known fusion approaches, we experimented with match score level fusion and decision level fusion. As one of the best match score level fusion methods, 'sum rule' and 'product rule' with 'min-max' normalization technique (Ross, Nandakumar, & Jain, 2006) have been applied. For decision level fusion approaches, 'AND' rule (Daugman, 2000), 'OR' rule (Daugman, 2000), 'majority voting'

Figure 8. Comparison between Markov chain-based rank fusion, fuzzy fusion, and match score fusion approaches

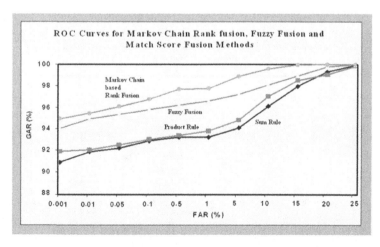

Figure 9. Comparison between Markov chain-based rank fusion, fuzzy fusion, and decision fusion approaches

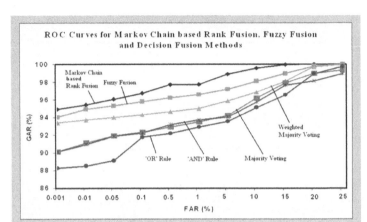

(Lam & Suen, 1997) and 'weighted majority voting' (Kuncheva, 2004) approaches have been applied. For the 'weighted majority voting' approaches, the highest weights have been assigned for irises and the lowest weights have been assigned for ears in the first virtual multimodal database. Figure 8 and Figure 9 show the outcomes of these experimentations.

From Figure 8 and Figure 9, it is clear that Markov chain based rank level fusion outperforms both match score level fusion and decision level fusion for the first virtual multimodal database. Among the match score level fusion methods,

'product rule' based method performs better than 'sum rule' based method. Among the decision level fusion approaches, 'weighted majority voting' method performs the best and the performance of 'OR' rule based method is the lowest. Also, in both experiments, fuzzy fusion method performs better than match score fusion and decision fusion approaches.

In order to evaluate the performance of the fuzzy fusion method, soft biometrics were used as additional information. The three soft biometrics: gender, ethnicity and eye color, were considered in this fuzzy fusion method. Figure 10 and Figure

Figure 10. Fuzzy fusion performance with the inclusion of soft biometric information tested with the first database

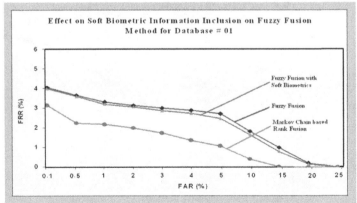

Figure 11. Fuzzy fusion performance with the inclusion of soft biometric information tested with the second database

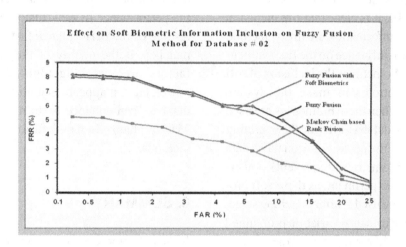

Figure 12. Comparison between EERs of different fusion approaches with the first datasets

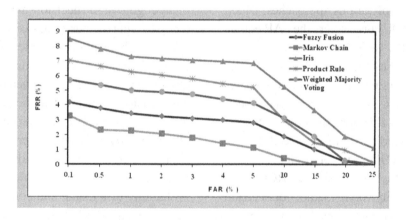

Figure 13. Comparison between EERs of different fusion approaches with the second datasets

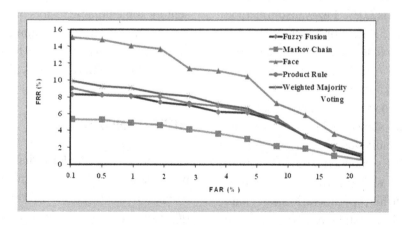

11 show the performances of these experimentations for the two previously introduced databases.

From Figure 10, it is clear that the inclusion of soft biometric information in fuzzy method does not have much influence on the final authentication outcome. Sometimes the inclusion of soft biometric information can make the system faster, especially, when these biometrics are used at first to divide the database (divide the operating spaces, i.e. the system operates only on those data where these soft biometrics are present). This is not the case in this experiment, as these soft biometric identifiers are used at after matching stage. Similar kind of performance variation is obtained with experiment utilizing the second database as shown in Figure 11.

Comparisons has also been made on the Equal Error Rates (EER) of various approaches, which is the point in a ROC curve where the FAR equals the FRR (1-GRR), with the two sets of virtual multimodal databases. Figure 12 shows the EERs comparison for the first virtual multimodal database. For this graph, ROC curve for iris has been used as a representative for unimodal matchers, ROC curve for Markov chain method as a rank fusion approach, ROC curve for product rule as a match score level fusion approach and ROC curve for weighted majority voting method for decision fusion approach as these are the best performing methods in their respective categories. From the curves, the best EER has been obtained for Markov chain method which is 2.03%. Fuzzy fusion method gives us an EER of 3.08%. For match score level fusion and decision fusion approaches, the EERs are 5.15% and 4.73%, respectively.

The same experimentations have been conducted on the second virtual multimodal database and the results are shown in Figure 13. As the qualities of these individual databases were not very good, the EERs obtained were not as high as EERs obtained with the first virtual multimodal database. In this experiment, Markov chain rank fusion method outperformed other approaches once again with an EER of 3.64%.

From all of these experiments, it is clear that the performances of individual matchers influence the performance of rank fusion or fuzzy fusion. As one of the performance factors of individual matchers is the quality of the database (others factors are matching algorithm, processing power, etc.), it appears that utilizing high quality database can enhance performances of both the Markov chain based rank fusion and fuzzy fusion methods.

6. SUMMARY

In this chapter, the fuzzy logic based fusion approach for multimodal biometric system has been described. It is a powerful intelligent tool used in many cognitive and decision-making systems. After discussing the basics of fuzzy logic, the fuzzy fusion mechanism in the context of a multimodal biometric system has been illustrated. A brief discussion on the research conducted for fuzzy logic based fusion in different application domains has also been presented. The system overview and the choice of fuzzy rules to govern the system have been presented. The biggest advantage of the system is that instead of binary Yes/No decision, the probability of a match and confidence level can now be obtained. Moreover, system can be easily adjusted by controlling weight assignment and fuzzy rules to fit changing conditions. After presenting some notable results of experimentation, the incorporation of soft biometric information with the fuzzy fusion method to make the system more accurate and robust has also been tested.

REFERENCES

Chaabane, L., & Abdelouahab, M. (2011). Improvement of brain tissue segmentation using information fusion approach. *International Journal of Advanced Computer Science and Applications*, *2*(6), 84–90.

Chang, K., Bowyer, K., & Barnabas, V. (2003). Comparison and combination of ear and face images in appearance-based biometrics. *IEEE Transactions on Pattern Analysis and Machine Intelligence*, 25, 1160–1165. doi:10.1109/TPAMI.2003.1227990.

Daugman, J. (2000). *Combining multiple biometrics*. Retrieved from http://www.cl.cam.ac.uk/users/jgd1000/combine/combine.html

Deng, Y., Su, X., Wang, D., & Li, Q. (2010). Target recognition based on fuzzy Dempster data fusion method. *Defence Science Journal*, 60(5), 525–530.

Harb, A. M., & Al-Smadi, I. (2006). Chaos control using fuzzy controllers (Mamdani model), integration of fuzzy logic and chaos theory. *Studies in Fuzziness and Soft Computing*, 187, 127–155. doi:10.1007/3-540-32502-6_6.

Kim, S.-W., Kim, K., Lee, J.-H., & Cho, D.-I. (2001). Application of fuzzy logic to vehicle classification algorithm in loop/piezo-sesor fusion systems. *Asian Journal of Control*, 3(2), 64–68.

Kuncheva, L. I. (2004). *Combining pattern classifiers: Methods and algorithms*. New York, NY: Wiley. doi:10.1002/0471660264.

Lam, L., & Suen, C. Y. (1997). Application of majority voting to pattern recognition: An analysis of its behavior and performance. *IEEE Transactions on Systems, Man, and Cybernetics. Part A, Systems and Humans*, 27(5), 553–568. doi:10.1109/3468.618255.

Monwar, M., Gavrilova, M., & Wang, Y. (2011). A novel fuzzy multimodal information fusion technology for human biometric traits identification. In *Proceedings of ICCI*CC*. Banff, Canada: IEEE.

Monwar, M. M., & Gavrilova, M. (2009). A multimodal biometric system using rank level fusion approach. *IEEE Transactions on Systems, Man, and Cybernetics. Part B, Cybernetics*, 39(4), 867–878. doi:10.1109/TSMCB.2008.2009071 PMID:19336340.

Monwar, M. M., & Gavrilova, M. (2011). Markov chain model for multimodal biometric rank fusion. In Proceedings of Signal, Image and Video Processing, (pp. 1-13). Springer-Verlag.

Pedrycz, W., & Gomide, F. A. C. (1998). *An introduction to fuzzy sets: Analysis and design complex adaptive systems*. Cambridge, MA: MIT Press.

Ross, A., Nandakumar, K., & Jain, A. K. (2006). *Handbook of multibiometrics*. New York, NY: Springer-Verlag.

Shafer, G. (1976). *A mathematical theory of evidence*. Princeton, NJ: Princeton University Press.

Solaiman, B., Pierce, L. E., & Ulaby, F. T. (1999). Multisensor data fusion using fuzzy concepts: Application to land-cover classification using ERS-1/JERS-1 SAR composites. *IEEE Transactions on Geoscience and Remote Sensing*, 37, 1316–1326. doi:10.1109/36.763295.

Wang, Y. (2009). Fuzzy inferences methodologies for cognitive informatics and computational intelligence. In *Proceedings of 8th IEEE International Conference on Cognitive Informatics*, (pp. 241-248). Hong Kong, China: IEEE.

Wang, Y. (2009). Toward a formal knowledge system theory and its cognitive informatics foundations. *IEEE Transactions of Computational Science*, 5, 1–19.

Wang, Y.-P., Dang, J.-W., Li, Q., & Li, S. (2007). Multimodal medical image fusion using fuzzy radial basis function neural networks. In *Proceedings of International Conference on Wavelet Analysis and Pattern Recognition*, (pp. 778-782). Beijing, China: IEEE.

Zadeh, L. A. (1965). Fuzzy sets. *Information and Control*, 8, 338–353. doi:10.1016/S0019-9958(65)90241-X.

Section 3
Applications in Security Systems

Chapter 8
Robotics and Multimodal Biometrics

ABSTRACT

This chapter presents a review on a new subfield of security research which transforms and expands the domain of biometrics beyond biological entities to include virtual reality entities, such as avatars, which are rapidly becoming a part of society. Artimetrics research at Cybersecurity Lab, University of Louisville, USA, and Biometric Technologies Lab, University of Calgary, Canada, builds on and expands such diverse fields of science as forensics, robotics, virtual worlds, computer graphics, biometrics, and security. Analyzing the visual properties and behavioral profiling can ensure verification and recognition of avatars. This chapter introduces a multimodal system for artificial entities recognition, simultaneously profiling multiple independent physical and behavioral characteristic of an entity, and creating a new generation multimodal system capable of authenticating both biological (human being) and non-biological (avatars) entities. At the end, this chapter focuses on some future research directions by discussing robotic biometrics beyond images and text-based communication to intelligent software agents that can emulate human intelligence. As artificial intelligence and virtual reality domains evolve, they will in turn give rise to new generation security solutions to identity management spanning both human and artificial entity worlds.

1. INTRODUCTION

Over the course of history, the greatest minds: scientists, philanthropists, educators, politicians, leaders, philosophers, were fascinated with the way human brain works. From Michelangelo to Lomonosov, from DaVinci to Einstein, there have been numerous attempts to uncover the mystery of human mind and to replicate its working first through simple mechanical devices an later, in the 20th century, through computing machines, software and robots.

DOI: 10.4018/978-1-4666-3646-0.ch008

In Alan Turing's 1950 work "Computing Machinery and Intelligence," Turing posed the question "can machines think?" In order to establish credible criteria to answer this question, he proposed a test, now known as "The Turing Test"—to evaluate a machine's ability to demonstrate intelligence. At the core of the test is conversation in a natural language between the human judge and the opponent, who can be either human or a machine. If the judge cannot reliably tell the machine from the human, the machine is said to have passed the test. In the light of recent developments, it can be viewed as the ultimate multimodal behavioral biometric, which can detect differences between a man and the machine.

Following Turing's work, another foundation of modern artificial intelligence was laid out by John von Neumann in the 1950s in his theory of automata and self-replicating machines. His theoretical concepts were based on those introduced by Alan Turing. The majority of research in this domain is focused on self-replicating programs and systems. Thus, development of computer viruses and spam e-mail applications has been quite fruitful. The obvious difficulty challenge lies with making robots reproduce themselves.

Self-replication in biological world is fairly well understood. The process of self-replication at the molecular level is responsible for all the living organisms on Earth today. Self-replication of non-biological entities is much less understood process. Much attention was devoted to the Cornell University researchers who have created a machine that can build copies of itself. Their robots are made up of a series of modular cubes—called "molecubes"—each identical to each other and each supplied with computer program for replication. The complete robot is built from a number of cubes, which connect by using electromagnets.

However, the bigger question of authentication and labeling of such "self-replicating" robots and software (such as viruses) has rarely been posed,

despite the growing concerns that uncontrollable development of self-replicating machines and machines with artificial intelligence can be somewhat harmful for the human society. And examples are numerous. Domestic and industrial robots, intelligent software agents, virtual world avatars and other artificial entities are quickly becoming a part of our everyday life. Just like it is necessary to be able to accurately authenticate identity of human beings, it is essential to be able to determine identity of the non-biological entities (Gavrilova & Yampolskiy, 2012). Military soldier-robots (Khurshid & Bing-Rong, 2004), robots museum guides (Charles, Rosenberg, & Thrun, 1999), software office assistants (Chen & Barthes, 2008), human-like biped robots (Lim & Takanishi, 2000), office robots (Asoh, Hayamizu, Hara, Motomura, Akaho, & Matsui, 1997), bots (Patel & Hexmoor, 2009), robots with human-like faces (Kobayashi & Hara, 1993), virtual world avatars (Tang, Fu, Tu, Hasegawa-Johnson, & Huang, 2008) and thousands of other man-made entities all have something in common: a pressing need for a decentralized, affordable, automatic, fast, secure, reliable, and accurate means of identity authentication. To address these concerns, the concept of *Artimetrics* – a field of study that will allow identifying, classifying and authenticating robots, software and virtual reality agents has been proposed in (Yampolskiy, 2007a; Yampolskiy & Govindaraju, 2007; Gavrilova & Yampolskiy, 2012).

While the area of robot and agent authentication may seem a bit futuristic at first, careful analysis of recent news stories shows that this is not quite so. To give just some examples: terrorists have been reported recruiting and communicating in virtual communities such as Second Life (Cole, 2008). Cybercrime, including identity theft, is rampant in virtual worlds populated by millions of avatars and operating multibillion dollar economies (Nood & Attema, 2009). Security experts have testified to the US Senate that defenses are lacking when it comes to emerging threats to the nation's Cyberinfrastructure. International teams of hackers assisted by semiautomatic hacking software agents have perpetrated numerous attacks against the Pentagon and other government agencies' computers and networks (Thompson, 2009).

A novel paradigm, unique to virtual communities, has appeared in recent years and was labeled "interreality." In the *Second Life* visitors are allowed to populate, build and exploit initially empty spaces. As a result "the new reality that is thus created is, remarkably enough not entirely 'virtual', but is becoming gradually more linked to our physical reality" (Nood & Attema, 2009). Relationships between social, economic, and psychological status of game players and their respected avatars in the virtual environment are a subject of current research. Early results show that avatars for the most parts resemble their "owners" rather than being completely virtual creations.

As the physical and the virtual worlds seem to come really close to each other, the distinction between the two begins to fade and the need arises for security systems capable of working in the contexts of interreality and augmented reality (Lyu, King, Wong, Yau, & Chan, 2005). In his dissertation 'Architecture of a Cyber Culture' published in 2003, Van Kokswijk describes this phenomenon as "the hybrid and absolute experience of physical and virtual reality". Interreality

is the creation of a hybrid total image of and in both the physical and virtual worlds. Unfortunately, currently available biometric systems are not designed to handle visual and behavioral variations observed in non-human agents and consequently perform extremely poorly if applied outside of their native domain (See Figure 1).

The question of security and identification of avatars in this "interreality" is an important one. Based on polls performed on Internet forums, users and members of on-line communities are not satisfied with the level of security in Second Life, with almost 40% of those polled asking for an additional security (Nood & Attema, 2009). Statistics is indeed alarming, and point to a larger issue which transpired from a human society onto the virtual worlds. More than half of the respondents report that they have been harassed (through imprisoning, stalking, gossiping, and using inappropriate language) and 40% indicate that certain actions should not be permitted in Second Life. Thus, the definite need in increasing and enforcing security is apparent, which motivates emerging research on security in the increasingly complex and interrelated virtual worlds.

We now further examine this rich research domain and make connections from cyber security area to multi-modal biometric.

Figure 1. Facial images of a humanoid robot-model, robot celebrity and a 3D-virtual avatar (Oh, Hanson, Kim, Han, Han, & Park, 2006; Ito, 2009; Oursler, Price, & Yampolskiy, 2009)

2. LITERATURE REVIEW

While automatic robot authentication or behavior analysis has not been closely investigated in literature, robot emotion recognition has been studies to some degree (Fong, Nourbakhsh, & Dautenhahn, 2003). In addition to experiments on understanding of emotional states of robots, some work has been started on general analysis of *avatar behavior*, such as the project on Avatar DNA. Together, the segments define the makeup of an avatar. The genes of the avatar are unique and include user biometric data, public key information, personal information, authentication information, creation data, etc. Verification modules in the virtual world collect information directly from the avatar to establish the roles and rights that should be granted to this user (Teijido, 2009).

In another experiment linking real world and the world of avatars, William Steptoe asked eleven volunteers some personal questions. During the interviews the volunteers were equipped with eye-tracking devices. A second group of volunteers watched videos of avatars as they communicated first group's answers. Some avatars had eye movements that mirrored those of the original volunteers, while others did not. The volunteers had to determine if the avatar was lying or telling the truth. Eye-movement seemed to have an impact on accurate detection of truthful statements, thus establishing the importance of body language in virtual world communication (Fisher, 2012).

Another example where behavioral biometrics developed for recognition of human beings is used in the recognition of intelligent software agents is given in (Yampolskiy & Govindaraju, 2008). It is considered in the context of using illegal bots in on-line game environments by players to obtain an unfair computer-assistant advantage over other players. A similar work on automatically distinguishing bots and humans apart was conducted in (Yampolskiy & Govindaraju, 2007). The research on spoofing of behavior-based biometric systems by artificially intelligent software agents was presented in (Yampolskiy, 2007b). The most comprehensive survey article to date, establishing theoretical groundwork for the research in authentication of non-biological entities, has appeared in 2012 and is partially based on BT Laboratory project with University of Louisville, USA (Gavrilova & Yampolskiy, 2012). In that work, over 200 references to all known to date biometric applications to robotics and virtual reality are given. Summary of most important developments presented in that survey is given below.

3. SURVEY OF NON-BIOLOGICAL ENTITIES

There are three main types of non-biological entities, that can be broadly classified as Virtual Beings (avatars), Intelligent Software Agents (bots), and Hardware Robots (Holz, Dragone, & O'Hare, 2009).

According to a dictionary, the word "Avatar" means: "embodiment: a new personification of a familiar idea"; or the manifestation of a Hindu deity (especially Vishnu) in human or superhuman or animal form. In an on-line community, Avatar is a virtual representation of a player in an on-line world, a software creation that exists in virtual environment but is controlled by a human player from the physical world. A comprehensive summary of avatar types is given in an on-line book by John Suler (2009). The book itself is not a typical publication – it exists only in the on-line form and evolves with time to reflect constant changes in virtual gaming communities. According to author of that book, the following types of avatars exist based on preferences and behavior of its human creator (Suler, 2009):

- **Odd/shocking Avatars:** Are unusual or strange;
- **Abstract Avatars:** Are represented by abstract art;

- **Billboard Avatars:** Are announcements or billboard notes;
- **Matching Avatars:** Usually appear together;
- **Clan Avatars:** Are associated with members of the same social group;
- **Animated Avatars:** Contain visual effects;
- *Animal* avatars are pets;
- **Cartoon Avatars:** Are based on cartoon drawings;
- **Celebrity Avatars:** Are associated with popular culture;
- **Real Face Avatars:** Are based on photographs of the actual users;
- **Idiosyncratic Avatars:** Are strongly related to a specific user;
- **Positional Avatars:** Are placed into a fixed location;
- **Power Avatars:** Channel authority.

Some of the ways to distinguish those avatars can be through their appearance, attributes, behavior, context, changes they undergo, and analyzing time dependencies. Thus, traditional image pattern recognition techniques and biometric behavioral methods can be successfully sued for this task. In avatars, behavior identification plays a key role to avatar's identity. Classification of their behavior can assist in understanding key identification trends. Thus, based on Suler (2009), such behavior can be expressed in Mischievous Pranks (such as smearing someone else's room, spoofing someone with "msay" command, or popping text balloon over someone's head), Flooding of the server by users who make rapid multiple changes of their avatars, Blocking (placing one's avatar on top or too close to another person's prop), Sleeping (by users who have walked away from their computer and their avatar fails to react), Eavesdropping (by reducing avatar to a single pixel and usernames to only one character, someone may become "invisible" and secretly listen in on conversations), Prop Dropping (placing an inappropriate or obscene prop in an empty room), Identity Disruption—people suffering from disturbances in their identity may act it out through frequently changing props they wear. Those behaviors resemble typical criminal behaviors of humans and thus require a high degree of attention from those in charge of security of the virtual communities.

Suler (2009) describes one such act. "Sometimes, it's hard even for sympathetic people to resist the antics and game-playing. One night, although trying to remain a neutral observer, I eventually found myself as an accomplice to another member in a prank where we set up an unmanned female prop in the spa pool. We used "msay" to talk through the prop while also talking to it as if it were another user. Essentially, it was a virtual ventriloquist act. Honey (the prop) was rather seductive towards the guests, and the guests all thought it was a "real" person. It was quite funny, although perhaps a bit mean to the poor naive guests who were unaware of the msay command." The quote above is very important as it introduced another virtual entity creation, or a "fake avatar", separate from legitimate avatars, that does not corresponds to a real person, but "appears" to be just like them and can sometimes fool even experienced users. Utilizing methods from biometric research as well as developing new approaches targeting specifically avatar authentication and behavior recognition can assist in identifying those "fake avatars" as well as classifying real ones.

4. AVATAR AUTHENTICATION

In this section, we first recall techniques for collecting and classifying databases of avatars and bots, described in Gavrilova and Yampolskiy (2012), moving on to propose a new way to synthesis the new images through application of biometric synthesis methods based on geometric processing and multi-resolution techniques. We then study the two main types of authentication in virtual world: visual and behavioral, and introduce the multi-modal system for enhanced performance.

4.1. Datasets Generation

As was pointed out before in this book, there are mainly three different types of biometric databases: true database, virtual database and synthetic database. For unimodal biometrics, there is an abundance of freely available datasets (Li & Jain, 2005), with vendor competitions often being held and benchmarks on recognition algorithm performance being established. This is not the case in the area of virtual reality. Labeled public datasets of avatar faces, robot faces, or attributed conversations from artificially intelligent agents are currently unavailable. Some recent papers attempted to tackle the problem by looking at applying methods for face generation (Klimpak, Grgic, & Delac, 2006; Gao, et al., 2008), gender attribution (Corney, Vel,, Anderson, & Mohay, 2002), and human versus bot classification to a virtual reality domain.

Prof Roman Yampolskiy have begun work on generation of a publicly available avatar face dataset (Oursler, Price, & Yampolskiy, 2009) by designing and implementing a scripting technique to automate the process of avatar face collection. Using the programming language AutoIt as well as a scripting language native to *Second Life*, better known as Linden Scripting Language (LSL), a successful generation of random avatars was achieved.

Figure 2 is a walkthrough of this process for the creation of randomly generated dataset of avatar faces.

The datasets generated by the scripted approach consists of ten pictures for each avatar taken from different angles. The images captured are in the Portable Network Graphics (PNG) format at a resolution of 1024 X 768 resulting in each image being between 110KB and 450KB in size. One upper body picture is taken as well as nine facial pictures, at various angles. These angles include the top, center, and bottom of the left, center and right side of each avatar's face. The images are named in a consistent format; stating the program,

Figure 2. Left: sample images for a robot-face dataset, currently limited to manual collection; right: automatically generated random avatar-faces (Oursler, Price, & Yampolskiy, 2009; Yampolskiy & Govindaraju, 2008)

gender, avatar number, and angle. The gender of the avatar is dependent upon the user's selection at the beginning of the process (Oursler, Price, & Yampolskiy, 2009).

While it is only possible to specify the desired amount of data, avatar's gender and overall area of knowledge about which intelligent agents communicate, it is still a challenge to generate data with specific characteristics. Some of the recently developed biometric synthesis processes can become handy in this task and described in the next section.

4.2. Synthetic Biometric and Artimetrics

A link between two areas—avatar generation and synthetic biometric generation, has largely remained unexplored until modern days. The early attempt to look at this problem was made in 1990s work, where authors discuss specific steps that need to be taken in order for avatar to be created automatically from a human face (Lyons, Plante, Jehan, Inoue, & Akamatsu, 1998). In addition, authors suggested that the process of avatar creation and authentication can be further enhanced by applying techniques from both biometric synthesis and biometric authentication.

Synthetic biometric is defined as "inverse problem of biometric" (Wang & Gavrilova, 2006) and is intended to create artificial phenomenon that does not exist in physical reality, but resembles it. The extensive research on synthetic biometric has been conducted at the Biometric Technologies Laboratory, University of Calgary, and results has been recently reported in the World Scientific book "Image Pattern Recognition: Synthesis and Analysis in Biometrics" (Yanushkevich, Gavrilova, Wang, & Srihari, 2007). In that study, link between biometric synthesis and inverse logic has been established, as the same principles can be applied to solve inverse logical problems and generate new synthetic biometric data.

There are numerous applications and high demand for new biometric databases to test new systems and study various phenomena, and many novel methods based on feature selection, pattern analysis, functional decomposition of spaces, signal processing, image decomposition and multi-resolution has been employed to generate new synthetic biometric data. A parallel between synthetic biometrics (such as fingerprints, irises, faces, ears, hands, behavioral trends and virtual bodies) and artificial entities, such as avatars, can be easily made. Both are created artificially, using computer technology and sophisticated algorithms, and both resemble human and human features. However, there are some substantial differences between synthetic biometrics and avatars.

Synthetic biometric, at least up to day, is completely non-personalized. It usually does not correspond to a single human or function, but possesses characteristics of multiple biometrics that were used in the process of new biometric entity synthesis. However, exactly this feature is most beneficial for new virtual dataset creation. Typically, data synthesis refers to the creation of new data to meet some intended purposes, and includes areas such as texture synthesis, domain specific rendering and biometric synthesis. One of the primary goals of the synthesis of biometric data is to provide databases for testing newly developed biometric algorithms (Luo & Gavrilova, 2006; Monwar & Gavrilova, 2006). For instance, in Yuan, Gavrilova, and Wang (2008) a new approach for facial synthesis and expression modeling based on the 2D and 3D mesh face models was created. Selection of control points in this method was advised by the three-dimensional Voronoi diagram.

However, this is often not the case for avatar generation. Its features, such as face, might be created based on a single source—a single photograph or a face drawing. The source in this case can be both real (actual photograph or a face scan) or artistically created (cartoon character, caricature etc.). It can resemble the source and have its own personally customized features, such as specified color of the eyes, selected hair style and accessories, voice, etc. Moreover, a concept of time can be introduced in order to create avatars with different age appearance.

While ability to customize the synthetic creation and to give avatars more individual features seems to be the case, there is still a very natural link between biometric synthesis studies and avatar creation and recognition domains. The latter is the core research topic of *artimetrics,* the area of studies focusing on artificial entities recognition in virtual reality domain using biometric principles. The area was created and popularized at the CyberSecurity lab at the University of Louisville and Biometric Technologies lab at the University of Calgary.

4.3. Visual Recognition

We now would like to concentrate specifically on visual recognition of avatars. Face recognition is the task performed by humans routinely and with an ease, and one of the first skills a child learns during his lifetime. Up to date, thousands of papers have been published on the topic, with comprehensive surveys of facial biometric research found in (Yang, Kriegman, & Ahuja, 2002; Zhao, Chellappa, Phillips, & Rosenfeld, 2003; Tan, Chen, Zhou, & Zhang, 2006) as well as in a recent book presenting state-of-the-art in facial biometric domain (Jain & Li, 2004). Knowledge based methods, such as the multi-resolution approach (Yang & Huang, 1994), capitalize on relationship between facial features. Feature invariant methods look for structures consistency under a variety of poses and lighting conditions. These methods are based on grouping of edges (Yow & Cipolla, 1997), space dependence matrix (Dai & Nakano, 1996), and Gausian models (McKenna, Gong, & Raja, 1998). Template matching extracts standard patterns of the face which are later compared to

regions being tested to determine the degree of correlation, classical examples include shape template (Craw, Tock, & Bennett, 1992) and Active Shape Model (ASM) (Lanitis, Taylor, & Cootes, 1995). Many popular appearance-based methods such as Eigenvector decomposition (Turk & Pentland, 1991), Support Vector Machines (SVM) (Osuna, Freund, & Girosi, 1997), Hidden Markov Model (HMM) (Rajagopalan, etc., 1998), Naïve Bayes Classifier (NBC) (Schneiderman & Kanade, 1998), and Neural Networks (NN) (Rowley, Baluja, & Kanade, 1998) learn facial templates from a set of training image.

It is interesting to observe that such adapted technique might prove more efficient in recognizing avatar faces rather than human faces. In a real photograph, there are many factors which can affect adversely the image quality and thus its recognition. Despite the process of normalization, the effects of air quality, lightning, reflections, person's posture, clothing, and possible movement, as well as the type of the physical medium used to capture the image (film, camera, cell phone) and the distance/positioning of this capturing device from the person, make the problem of face recognition highly difficult to tackle. On contrary, in the virtual world, while some variability exists, the nature of the avatar being a computer generated entity makes it much easier to perform matching. An example of application of feature-based (geometry-based) method to avatar recognition is given in Figure 3 (Gavrilova & Yampolskiy, 2012). The method is similar to geometry-based recognition method for real human faces. Another example in Figure 4 (Mohamed & Gavrilova, 2012) demonstrates that color and texture information can be also taken into an account. Here, wavelet methodology is used to iteratively subdivide the image and extract appearance-based features.

Another important factor to consider is that in some cases there is a strong resemblance of avatar to its human creator, which makes it possible to use the results of successful avatar recognition

Figure 3. Feature-based facial recognition applied to an avatar's face (Gavrilova & Yampolskiy, 2012)

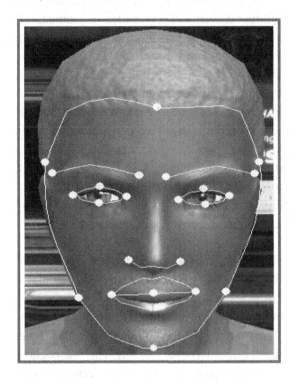

for human recognition, and vice versa. This will, in turn, open a new area of *virtual world biometric*, or augmenting the actual biometric with results of recognition in virtual worlds.

4.4. Behavioral Authentication

Facial recognition, facial expression analysis and facial synthesis are highly prominent and actively researched areas of biometric (Monwar & Gavrilova, 2008; Jain & Li, 2004). Facial recognition is used in the areas such as medical science, lie detection, HCI, pain analysis, crowd control, education, on-line games. Facial expression analysis has been recently combined with EEG (Electroencephalogram) studies of human brain activities under an impact of outside influences. Emotion recognition has been actively researched in game and movie industry, with both photorealistic and non-photorealistic rendering methods used to animate characters. Finally, facial synthesis is used for validating real-life scenarios in training

Figure 4. An example of an avatar appearance-based face analysis (Mohamed & Gavrilova, 2012)

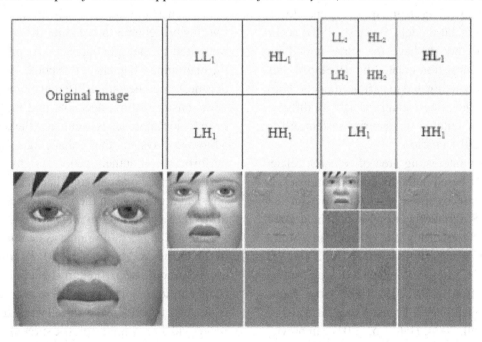

awareness systems personnel, and as a reasonably inexpensive tool to test sophisticated algorithms on massive datasets of synthetic data.

One of emerging recent trends is combining emotion recognition with facial recognition, another—in using supplementary metadata to enhance recognition capabilities of the system.

One approach to the problem takes into an account subtle expression changes and performs morphing expression images in 2D and 3D based on the powerful computational geometry methods (Wang, Gavrilova, Luo, & Rokne, 2006; Wecker, Samavati, & Gavrilova, 2005; Yuan, Gavrilova, & Wang, 2008). This work makes a number of important contributions to the field of expression modeling and morphing: it is one of the first applications of the sketch-based approach to facial image generation and the first one that preserves and utilizes subtle expression lines. It provides a simple fully automated algorithm based on the distance transform that computed the mapping between pixels in a monochrome image through a clever process of sweeping the image and reusing the information obtained on the previous step. It also provides a combination of Sibson coordinates and Delaunay triangulation mesh to generate and morph 3D facial models. Because all the generated facial models have the same underlying structure, animation created by developed tools can be easily retargeted to various models. Thus, it allows generating facial models with different expressions suitable for security system testing in real-world scenarios.

Another interesting area of research related to behavior-based authentication if forensics, or authorship recognition. Analysis of plain text (Juola & Sofko, 2004; Koppel & Schler, 2004; Koppel, Schler, & Mughaz, 2004), emails (Stolfo, Hershkop, Wang, Nimeskern, & Hu, 2003; Vel, Anderson, Corney, & Mohay, 2001), and source code (Spafford & Weeber, 1992; Gray, Sallis, & MacDonell, 1997; Frantzeskou, Gritzalis, & MacDonnell, 2004; Halteren, 2004) can provide

insights on specific use of terms and unique writing/typing style, which can be as handy in virtual worlds as in real ones. The list of commonly used text descriptors is provided in (Stamatatos, Fakotakis, & Kokkinakis, 1999). Once linguistic features have been established, many learning methods can be used. Support Vector Machines (Joachim, Jorg, Edda, & Paass, 2003), Bayesian classifiers (Kjell, 1994), statistical analysis methods (Stamatatos, Fakotakis, & Kokkinakis, 2001) can learn common patterns and assist in establishing the authorship of the text. Clearly, those methods can be applied almost directly to behavior pattern analysis avatars based on the way they communicate in the virtual world.

4.5. Multi-Modal Biometrics and Robot/Avatar Recognition

Biometric system based solely on a single biometric may not always identify the artificial entity in the most optimal or precise way. This is especially true in the presence of complex patterns, conflicting behavior, abnormal or noisy data, intentional or accidental mischief etc. A reliable and successful multibiometric system normally utilizes an effective fusion scheme to combine the information presented by multiple matchers. As presented in the beginning Chapters of this book, over the last decade, researchers tried different biometric traits with sensor, feature, decision, and match score level fusion approaches to enhance the security of a biometric system, thus enhancing security and performance of authentication system.

In a similar manner, together behavioral and physical artimetrics in the virtual worlds can be utilized as a part of a *Physoemotional Artimetric* system which is a multimodal system introduced in (Gavrilova & Yampolskiy, 2012). In fact, this approach could be particularly beneficial for artificial entities recognition as there are more ways to disguise yourself in the virtual world than in the real world. For example, plastic surgery to change someone's appearance might be not as common or popular procedure in real world as changing someone's appearance in a virtual one. However, it is not that easy to change/modify someone's behavior, customs, or manner of speech in both real and virtual environment. Thus, a reverse dependence becomes true for virtual world: behavioral patterns might have a higher influence and imply a higher recognition rates than appearance-based ones. While appearance recognition value somewhat diminishes, the behavior pattern study, popular activities analysis, social surroundings, manner of speech, favorite places to visit, hobbies, skills, art and even wealth in virtual world can supply the crucial information for virtual entity authentication. Combining the visual and behavioral artimetrics using multi-modal biometric approach is emerging area of research that was introduced in (Gavrilova & Yampolskiy, 2012). In the same work, another highly original notion of Multi-Dimensional system was proposed. Such system is a merge between virtual and real worlds, which empowers authentication of avatar through its creator authentication, and vice versa, in both virtual and real domains.

5. APPLICATIONS

Cyber security, without a doubt, is one of the key concerns of many modern organizations as well as private citizens worldwide.

There have been a number of attempts by malicious intelligent software to obtain unlawful access to information or system resources. It affects security of virtual communities, social networks, and government supported cyber-infrastructures. Employing new methods to counter those threats is one of the goals of biometric and cyber security.

Extensive research on behavior based profiling of software agents as an unobtrusive way of separating helpful bots from malware (Gavrilova

& Yampolskiy, 2012). Additional research in artimetrics is likely to produce more behavior-profiling methods specifically designed to take advantage of the unique structure of artificially intelligent programs (Ahn, Blum, Hopper, & Langford, 2003; Baird & Bentley, 2005; Bentley & Mallows, 2006; Misra & Gaj, 2006; Yampolskiy & Govindaraju, 2007).

Exploring similarities between avatar and avatar owner behavior and their mutual behavioral profiling is highly likely to help to understand specific behavioral trends of both. Moreover, any changes in their behavior might suggest a higher risk of possible security attack, or simply pinpoint a difference between normal and abnormal behavior (Gavrilova & Yampolskiy, 2012).

Using novel research in EEG (Electroencephalogram) allows to further relate human and avatar emotions in an attempt to understand how outside factors (smell, sound, touch) impact human emotions and how this information can be expressed by avatars in virtual worlds (Sourina, Sourin, & Kulish, 2009).

Securing interaction between botnents, groups of intelligent cooperating agent and mixed robot/human teams are quickly emerging in numerous domains. Different pieces of intelligent software or between a human being and an instance of intelligent software in a virtual world is also an important domain (Kanda, Ishiguro, Ono, Imai, & Mase, 2002; Klingspor, Demiris, & Kaiser, 1997).

New applications include detecting cheating in games based on assistance from artificial intelligence software provide visual and behavioral search capabilities for the virtual worlds, such as Second Life, and target merchandise marketing in virtual worlds only to agents matching a certain profile (Gavrilova & Yampolskiy, 2012).

While using digital signatures and encrypted communication became a norm in web-based interconnected environment where humans operate, development of similar protocols for identifying robots and intelligent agents based on biometric security ideas is a novel approach to identity management.

Other important applications of artimetrics can be found in banking, border control, immigration policies, e-commerce, virtual community development, social networks, on-line game communities and any other collaborative environment.

6. SUMMARY AND FUTURE WORK

This review chapter describes a new subfield of security research which transforms and expands the domain of biometrics beyond biological entities to include virtual reality entities, such as avatars, which are rapidly becoming a part of society. Artimetrics research at Cybersecurity Lab, University of Louisville, USA and Biometric Technologies Lab, University of Calgary, Canada, builds on and expands such diverse fields of science as forensics, robotics, virtual worlds, computer graphics, biometrics and security. The chapter discusses how verification and recognition of avatars can be ensured by analyzing their visual properties and behavioral profiling. It also introduces a multimodal system for artificial entities recognition, simultaneously profiling multiple independent physical and behavioral characteristic of an entity, and creating a new generation multimodal system capable of authenticating both biological (human being) and non-biological (avatars) entities.

Future cyber security and biometric research may involve in-depth study of other visual and behavioral approaches to avatar/robot security based on introduction of new characteristics and abilities in the intelligent artificial entities of tomorrow. The modern day research is capable of expanding robotic biometrics beyond images and text-based communication to intelligent software agents that emulate quite successfully human intelligence. As artificial intelligence and virtual reality domains evolve, it will in turn give rise

to new generation security solutions to identity management spanning both human and artificial entity worlds.

REFERENCES

Ahn, L. V., Blum, M., Hopper, N., & Langford, J. (2003). CAPTCHA: Using hard AI problems for security. In Proceedings of Eurocrypt, 2003. Eurocrypt.

Asoh, H., Hayamizu, S., Hara, I., Motomura, Y., Akaho, S., & Matsui, T. (1997). Socially embedded learning of the office-conversant mobile robotjijo-2. In *Proceedings of 15th International Joint Conference on Artificial Intelligence*. ACM/IEEE.

Baird, H. S., & Bentley, J. L. (2005). Implicit CAPTCHAs. In Proceedings *of the SPIE/IS&T Conference on Document Recognition and Retrieval XII (DR&R2005)*. San Jose, CA: SPIE/IS&T.

Bentley, J., & Mallows, C. L. (2006). CAPTCHA challenge strings: Problems and improvements. In *Proceedings of Document Recognition & Retrieval*. IEEE.

Boyd, R. S. (2010). *Feds thinking outside the box to plug intelligence gaps*. Retrieved from http://www.mcclatchydc.com/2010/03/29/91280/feds-thinking-outside-the-box.html

Charles, J. S., Rosenberg, C., & Thrun, S. (1999). Spontaneous, short-term interaction with mobile robots. In *Proceedings of IEEE International Conference on Robotics and Automation*, (pp. 658-663). IEEE.

Chen, K.-J., & Barthes, J.-P. (2008). Giving an office assistant agent a memory mechanism. In *Proceedings of 7th IEEE International Conference on Cognitive Informatics*, (pp. 402–410). IEEE.

Cole, J. (2008). *Second life salon.* Retrieved from http://www.salon.com/opinion/feature/2008/02/25/avatars/

Corney, M., Vel, O. D., Anderson, A., & Mohay, G. (2002). Gender-preferential text mining of e-mail discourse. In *Proceedings of 18th Annual Computer Security Applications Conference*, (pp. 282-289). Brisbane, Australia: IEEE.

Craw, I., Tock, D., & Bennett, A. (1992). Finding face features. In *Proceedings of Second European Conference on Computer Vision*, (pp. 92-96). Santa Margherita Ligure, Italy: IEEE.

Dai, Y., & Nakano, Y. (1996). Face-texture model based on SGLD and its application in face detection in a color scene. *Pattern Recognition, 29*(6), 1007–1017. doi:10.1016/0031-3203(95)00139-5.

Fisher, R. (2012). Avatars can't hide your lying eyes. *New Scientist.* Retrieved from www.newscientist.com/article/mg20627555.600-avatars-cant-hide-your-lying-eyes.html

Fong, T. W., Nourbakhsh, I., & Dautenhahn, K. (2003). A survey of socially interactive robots. *Robotics and Autonomous Systems, 42*, 143–166. doi:10.1016/S0921-8890(02)00372-X.

Frantzeskou, G., Gritzalis, S., & MacDonell, S. (2004). Source code authorship analysis for supporting the cybercrime investigation process. In *Proceedings of 1st International Conference on eBusiness and Telecommunication Networks - Security and Reliability in Information Systems and Networks Track*, (pp. 85-92). Setubal, Portugal: IEEE.

Gao, W., Cao, B., Shan, S., Chen, X., Zhou, D., Zhang, X., & Zhao, D. (2008). The CAS-PEAL large-scale Chinese face database and baseline evaluations. *IEEE Transactions on Systems, Man and Cybernetics. Part A, 38*(1), 149–161.

Gavrilova, M., & Yampolskiy, R. (2012). Applying biometric principles to avatar recognition. In *Proceedings of IEEE RAM.* IEEE.

Gianvecchio, S., Xie, M., Wu, Z., & Wang, H. (2008). Measurement and classification of humans and bots in internet chat. In *Proceedings of 17th Conference on Security Symposium*, (pp. 155-169). San Jose, CA: IEEE.

Gray, A., Sallis, P., & MacDonell, S. (1997). Software forensics: Extending authorship analysis techniques to computer programs. In *Proceedings of 3rd Biannual Conference of the International Association of Forensic Linguists.* IEEE.

Halteren, H. V. (2004). Linguistic profiling for author recognition and verification. In *Proceedings of ACL.* ACL.

Holz, T., Dragone, M., & O'Hare, G. P. (2009). Where robots and virtual agents meet: A survey of social interaction research across Milgram's reality-virtuality continuum. *International Journal of Social Robotics, 1*(1). doi:10.1007/s12369-008-0002-2.

Ito, J. (2009). Fashion robot to hit Japan catwalk. *PHYSorg.* Retrieved from www.physorg.com/pdf156406932.pdf.

Jain, A. K., & Li, S. Z. (2004). *Handbook on face recognition.* New York, NY: Springer-Verlag.

Joachim, D., Jorg, K., Edda, L., & Paass, G. (2003). Authorship attribution with support vector machines. In *Proceedings of Applied Intelligence* (pp. 109–123). IEEE.

Juola, P., & Sofko, J. (2004). Proving and improving authorship attribution. In *Proceedings of CaSTA.* CaSTA.

Kanda, T., Ishiguro, H., Ono, T., Imai, M., & Mase, K. (2002). Multi-robot cooperation for human-robot communication. In *Proceedings of 11th IEEE International Workshop on Robot and Human Interactive Communication*, (pp. 271- 276). IEEE.

Khurshid, J., & Bing-Rong, H. (2004). Military robots - A glimpse from today and tomorrow. In *Proceedings of 8th Control, Automation, Robotics and Vision Conference*, (pp. 771-777). IEEE.

Kjell, B. (1994). Authorship attribution of text samples using neural networks and Bayesian classifiers. In *Proceedings of, IEEE International Conference on Systems, Man, and Cybernetics. 'Humans, Information and Technology'*, (pp. 660-1664). San Antonio, TX: IEEE.

Klimpak, B., Grgic, M., & Delac, K. (2006). Acquisition of a face database for video surveillance research. In *Proceedings of 48th International Symposium focused on Multimedia Signal Processing and Communications*, (pp. 111-114). IEEE.

Klingspor, V., Demiris, J., & Kaiser, M. (1997). Human-robot-communication and machine learning. *Applied Artificial Intelligence, 11*, 719–746.

Kobayashi, H., & Hara, F. (1993). Study on face robot for active human interface-mechanisms of facerobot and expression of 6 basic facial expressions. In *Proceedings of 2nd IEEE International Workshop on Robot and Human Communication*, (pp. 276-281). Tokyo, Japan: IEEE.

Koppel, M., & Schler, J. (2004). Authorship verification as a one-class classification problem. In *Proceedings of 21st International Conference on Machine Learning*, (pp. 489-495). Banff, Canada: IEEE.

Koppel, M., Schler, J., & Mughaz, D. (2004). Text categorization for authorship verification. In *Proceedings of Eighth International Symposium on Artificial Intelligence and Mathematics*. Fort Lauderdale, FL: IEEE.

Lanitis, A., Taylor, C. J., & Cootes, T. F. (1995). An automatic face identification system using flexible appearance model. *Image and Vision Computing, 13*(5), 393–401. doi:10.1016/0262-8856(95)99726-H.

Li, S., & Jain, A. K. (2005). *Handbook of face recognition - Face databases*. New York, NY: Springer.

Lim, H.-O., & Takanishi, A. (2000). Waseda biped humanoid robots realizing human-like motion. In *Proceedings of 6th International Workshop on Advanced Motion Control*, (pp. 525-530). Nagoya, Japan: IEEE.

Luo, Y., & Gavrilova, M. (2006). 3D facial model synthesis using voronoi approach. In *Proceedings of IEEE ISVD*, (pp. 132-137). Banff, Canada: IEEE.

Lyons, M., Plante, A., Jehan, S., Inoue, S., & Akamatsu, S. (1998). Avatar creation using automatic face recognition. In *Proceedings of ACM Multimedia 98*, (pp. 427-434). Bristol, UK: ACM.

Lyu, M. R., King, I., Wong, T. T., Yau, E., & Chan, P. W. (2005). ARCADE: Augmented reality computing arena for digital entertainment. In *Proceedings of IEEE Aerospace Conference*. Big Sky, MT: IEEE.

McKenna, S., Gong, S., & Raja, Y. (1998). Modelling facial colour and identity with Gaussian mixtures. *Pattern Recognition, 31*, 1883–1892. doi:10.1016/S0031-3203(98)00066-1.

Misra, D., & Gaj, K. (2006). Face recognition CAPTCHAs. In *Proceedings of International Conference on Telecommunications, Internet and Web Applications and Services*. IEEE.

Mohamed, A., Gavrilova, A., & Yampolskii, R. (2012). Artificial face recognition using wavelet adaptive LBP with directional statistical features. In *Proceedings of CyberWorlds 2012*. IEEE. doi:10.1109/CW.2012.11.

Monwar, M. M., & Gavrilova, M. (2008). FES: A system of combining face, ear and signature biometrics using rank level fusion. In *Proceedings of 5th International Conference on Information Technology: New Generations*, (pp. 922-927). Las Vegas, NV: IEEE.

Nood, D. D., & Attema, J. (2009). *The second life of virtual reality*. Retrieved from http://www.epn.net/interrealiteit/EPN-REPORT-The_Second_Life_of_VR.pdf

Oh, J.-H., Hanson, D., Kim, W.-S., Han, I. Y., Han, Y., & Park, I.-W. (Eds.). (2006). *Proceedings of international conference on intelligent robots and systems*. Daejeon, South Korea: IEEE.

Osuna, E., Freund, R., & Girosi, F. (1997). Training support vector machines: An application to face detection. In *Proceedings of IEEE Conference on Computer Vision and Pattern Recognition*, (pp. 130-136). IEEE.

Oursler, J. N., Price, M., & Yampolskiy, R. V. (2009). Parameterized generation of avatar face dataset. In *Proceedings of 14th International Conference on Computer Games: AI, Animation, Mobile, Interactive Multimedia, Educational & Serious Games*. Louisville, KY: IEEE.

Patel, P., & Hexmoor, H. (2009). Designing BOTs with BDI agents. In *Proceedings of International Symposium on Collaborative Technologies and Systems (CTS)*. (pp. 180-186). Carbondale, PA: IEEE.

Rajagopalan, A., Kumar, K., Karlekar, J., Manivasakan, R., Patil, M., & Desai, U. … Chaudhuri, S. (1998). Finding faces in photographs. In *Proceedings of, 6th IEEE Intern. Conference on Computer Vision*, (pp. 640-645). IEEE.

Ross, A. (2007). An introduction to multibiometrics. In *Proceedings of 15th European Signal Processing Conference*. Poznan, Poland: IEEE.

Ross, A., & Jain, A. K. (2003). Information fusion in biometrics. *Pattern Recognition Letters, 24*, 2115–2125. doi:10.1016/S0167-8655(03)00079-5.

Rowley, H., Baluja, S., & Kanade, T. (1998). Neural network-based face detection. *IEEE Transactions on Pattern Analysis and Machine Intelligence, 20*(1), 23–38. doi:10.1109/34.655647.

Schneiderman, H., & Kanade, T. (1998). Probabilistic modeling of local appearance and spatial relationships for object recognition. In *Proceedings of IEEE Conference on Computer Vision and Pattern Recognition*, (pp. 45-51). IEEE.

Sourina, O., Sourin, A., & Kulish, V. (2009). EEG data driven animation and its application. [MIRAGE.]. *Proceedings of MIRAGE, 2009*, 380–388.

Spafford, E. H., & Weeber, S. A. (1992). Software forensics: Can we track code to its authors? In *Proceedings of 15th National Computer Security Conference*, (pp. 641-650). IEEE.

Stamatatos, E., Fakotakis, N., & Kokkinakis, G. (1999). Automatic authorship attribution. In *Proceedings of Ninth Conference of the European Chapter of the Association of Computational Linguistics*, (pp. 158-164). Bergen, Norway: IEEE.

Stamatatos, E., Fakotakis, N., & Kokkinakis, G. (2001). Computer-based authorship attribution without lexical measures. *Computers and the Humanities, 35*(2), 193–214. doi:10.1023/A:1002681919510.

Stolfo, S. J., Hershkop, S., Wang, K., Nimeskern, O., & Hu, C.-W. (2003). A behavior-based approach to securing email systems. *Mathematical Methods. Models and Architectures for Computer Networks Security, 2776*, 57–81. doi:10.1007/978-3-540-45215-7_5.

Suler, J. (2009). *The psychology of cyberspace.* Retrieved from http://psycyber.blogspot.com

Tan, X., Chen, S., Zhou, Z.-H., & Zhang, F. (2006). Face recognition from a single image per person: A survey. *Pattern Recognition, 39*(9), 1725–1745. doi:10.1016/j.patcog.2006.03.013.

Tang, H., Fu, Y., Tu, J., Hasegawa-Johnson, M., & Huang, T. S. (2008). Humanoid audio-visual avatar with emotive text-to-speech synthesis. *IEEE Transactions on Multimedia, 10*, 969–981. doi:10.1109/TMM.2008.2001355.

Teijido, D. (2009). Information assurance in a virtual world. In *Proceedings of Australasian Telecommunications Networks and Applications Conference.* Canberra, Australia: IEEE.

Thompson, B. G. (2009). *The state of homeland security.* Retrieved from http://hsc-democrats.house. gov/SiteDocuments/20060814122421-06109.pdf

Turk, M., & Pentland, A. (1991). Eigenfaces for recognition. *Journal of Cognitive Neuroscience, 3*(1), 71–86. doi:10.1162/jocn.1991.3.1.71.

Vel, O. D., Anderson, A., Corney, M., & Mohay, G. (2001). Mining email content for author identification forensics. *SIGMOD Record, 30*(4), 55–64. doi:10.1145/604264.604272.

Wang, C., & Gavrilova, M. (2006). Delaunay triangulation algorithm for fingerprint matching. In *Proceedings of ISVD*, (pp. 208-216). Banff, Canada: ISVD.

Wang, H., Gavrilova, M., Luo, Y., & Rokne, J. (2006). An efficient algorithm for fingerprint matching. In *Proceedings of International Conference on Pattern Recognition*, (pp. 1034-1037). IEEE.

Wecker, L., Samavati, F., & Gavrilova, M. (2005). Iris synthesis: A multi-resolution approach. In *Proceedings of 3rd International Conference on Computer Graphics and Interactive Techniques in Australasia and South East Asia*, (pp. 121-125). IEEE.

Yampolskiy, R. V. (2007). Behavioral biometrics for verification and recognition of AI programs. In *Proceedings of 20th Annual Computer Science and Engineering Graduate Conference (GradConf).* Buffalo, NY: GradConf.

Yampolskiy, R. V. (2007). Mimicry attack on strategy-based behavioral biometric. In *Proceedings of 5th International Conference on Information Technology: New Generations*, (pp. 916-921). Las Vegas, NV: IEEE.

Yampolskiy, R. V., & Govindaraju, V. (2007). Behavioral biometrics for recognition and verification of game bots. In *Proceedings of the 8th Annual European Game-On Conference on simulation and AI in Computer Games.* Bologna, Italy: IEEE.

Yampolskiy, R. V., & Govindaraju, V. (2007). Embedded non-interactive continuous bot detection. *ACM Computers in Entertainment, 5*(4), 1–11. doi:10.1145/1324198.1324205.

Yampolskiy, R. V., & Govindaraju, V. (2008). Behavioral biometrics for verification and recognition of malicious software agents. In *Proceedings of SPIE Defense and Security Symposium.* Orlando, FL: IEEE.

Yang, G., & Huang, T. S. (1994). Human face detection in complex background. *Pattern Recognition*, *27*(1), 53–63. doi:10.1016/0031-3203(94)90017-5.

Yang, M.-H., Kriegman, D. J., & Ahuja, N. (2002). Detecting faces in images: A survey. *IEEE Transactions on Pattern Analysis and Machine Intelligence*, *24*(1).

Yanushkevich, S., Gavrilova, M., Wang, P., & Srihari, S. (2007). *Image pattern recognition: Synthesis and analysis in biometrics*. New York, NY: World Scientific Publishers.

Yow, K. C., & Cipolla, R. (1997). Feature-based human face detection. *Image and Vision Computing*, *15*(9), 713–735. doi:10.1016/S0262-8856(97)00003-6.

Yuan, L., Gavrilova, M., & Wang, P. (2008). Facial metamorphosis using geometrical methods for biometric applications. *International Journal of Pattern Recognition and Artificial Intelligence*, *22*(3), 555–584. doi:10.1142/S0218001408006399.

Zhao, W., Chellappa, R., Phillips, P. J., & Rosenfeld, A. (2003). Face recognition: A literature survey. *ACM Computing Surveys*, *35*(4), 399–458. doi:10.1145/954339.954342.

Chapter 9
Chaotic Neural Networks and Multi-Modal Biometrics

ABSTRACT

Neural network is a collection of interconnected neurons with the ability to derive conclusion from imprecise data that can be used to both identify and learn patterns. This chapter presents the concept of neural network as an intelligent learning tool for biometric security systems. Neural networks have been extensively used in a variety of computational and optimization problems. In the first half of this chapter, focus is given to a specific topic—chaos in neural network. A detailed description of an on-demand chaotic noise injection method recently developed to deal with a common drawback of non-autonomous methods—their blind noise injecting strategy—is presented. The second part of the chapter discusses the issue of high-dimensionality in the context of a complex biometric security system. The amount of data and its complexity can be overwhelming, and one way of dealing with this issue is to use the dimensionality reduction techniques, which are typically based on clustering or transformations from one space to another. The reduced dimensionality vector can be then used in the energy model for an associative memory, which will learn the data patterns. The benefit is that this is a learner system that converges the given set of vectors to the stored pattern in a network, which can be later used for biometric recognition and also for identifying the most significant biometric patterns. At the end of this chapter, some examples are presented showing the feasibility of using such approach in biometric domain—both for single and multi-modal biometric.

DOI: 10.4018/978-1-4666-3646-0.ch009

1. INTRODUCTION

In the previous chapters, we exploited notion of multi-modal biometrics, specifically ranked-level fusion approach to design of highly reliable and accurate biometric system. We also looked at advantages which can be gained by utilizing additional information about a subject, such as height, age and gender, or so-called soft biometric patterns. We also looked at the new research domain combining behavioral and appearance-based characteristics one example of artificial entities, such as robots, avatars and intelligent software agents.

In this section, we look at learning approaches and try to bring benefits of utilizing and identifying most prominent/significant patterns in multitude of biometric data, which not necessary originates from the same biometric. Features from different biometric sources can be combined at either before-matching or after-matching stages and then most significant traits can be identified through dimensionality reduction or adaptive learning approaches.

The proposed methodology has a number of interesting features:

- There is no dependency on one strong biometric. Even if the sample is not available/corrupted, accuracy of recognition and performance of security system will not be compromised;
- The combining of traits can be done on both pre-matching or post-matching levels, thus allowing choice for system design and implementation/integration in real applications;
- Computational complexity can be controlled through clever dimensionality reduction techniques, thus allowing for real-time performance;
- Utilizing learning approaches can lead to better recognition rates as training will take place on biometric database prior to matching.

In a practical biometric security system (i.e. a system that employs biometrics for personal recognition), there are a number of important issues that should be considered, including *performance* (achievable recognition accuracy and speed) and *circumvention* (system resistance to noise and to being fooled by fraudulent methods). In order for biometric system to meet the requirements on performance and circumvention, more than one type of biometric is required. Hence, the need arises for the use of multi-modal biometrics, which is a combination of different biometric recognition technologies, varying from physical biometrics (such as face, iris, and fingerprint recognition) to behavioral characteristics (i.e. signature, voice, and gate).

Having to deal with different biometrics characteristics and specifications usually leads to a number of issues that should be addressed in a multi-modal biometric system (Dalenol, Dellisanti, & Giannini, 2008; Ho, Hull, & Srihari, 1994; Johnson, 1991; Verlinde & Cholet, 1999). In such a system, one of the common problems is the high dimensionality of the data which negatively impacts the security system performance. Hence, dimensionality reduction methodologies need to be used. However, they have not been considered in recent multi-modal biometric systems due to gap between recently developed dimensionality reduction techniques in data mining and data analysis of biometric features. To correct this situation, a methodology for reducing the search space of all possible subspaces by utilizing axis-parallel subspaces and clustering can be used. This also helps in dealing with noisy data and makes the biometric system more error-proof.

Applying learning method based on chaotic neural network can help to improve the performance and circumvention of multi-modal biometric system.

Any multi-biometric system should be able to perform efficiently under a variety of conditions. Both data and algorithms operating on it play a key role in ensuring system resistance to noise, sample quality, or even a complete lack of some biometric features. A key problem to address is the high dimensionality of data in order to achieve the above goal. First of all, reducing data dimensionality in a biometric system model makes it easier to identify data patterns, and thus improves speed and reliability of a biometric verification system. Second, the noise, distortion, illumination, template aging and other artifacts are easier to deal with. Third, as part of the template matching process, the matching distance values from selected individual biometrics has to be analyzed. These values are difficult to work with since their amount can be as high as millions, and hence dimensionality reduction methods must be used. Thus, it is necessary to find a way to work with high dimensional data and create a new data subspace clustering/classification to meet the goals outlined above.

There are plenty of methods developed to combat high-dimensionality problem. Clustering and projection of a multi-dimensional vector to a lower spaces is one of approaches (Achtert, Bohm, Kriegel, Kroger, & Zimek, 2007; Achtert, Bohm, David, Kroger, & Zimek, 2008). Using locally adaptive methods for large databases is another (Chakrabarti, Keogh, Mehrotra, & Pazzani, 2002). Dealing with uncertainty in data often calls for additional attention to details while clustering (Charu, Aggarwal, & Philip, 2008). Special methods are also needed for dealing with spatial databases (Ester, Kriegel, Sander, & Xu, 1996). Feature selection methods such as Principal Component Analysis (PCA) can be used to map the original data space to a lower-dimensional data space, where the points may cluster better

and the resulting clusters may be more meaningful (Jain, Ross, & Pranbhakar, 2004). Even though PCA has been successfully used in face biometric research, unfortunately, such feature selection or dimensionality reduction techniques is not suitable to be applied to biometric clustering problems.

2. SYSTEM ARCHITECTURE

To further understand the main benefits of this idea, it is important to understand the general architecture of the neural network based biometric recognition system. An illustration of the proposed multi-modal biometric system can be viewed in Figures 1 and 2. Figure 1 presents a flowchart for creating templates for training biometric system dataset. For each of system N users, their individual biometrics are being collected and then represented as a reduced-dimensionality feature vector set. This feature vector set will be then given as an input to neural network based on chaotic associative memory to learn common patterns for subsequent user recognition.

Figure 2 shows the process of a new user enrollment and matching. A new user must initially be enrolled before using the system for authentication. This process must be monitored by an authorized person who enrolls the new user. This authorized person has to control that the enrollee is who he or she claims really to be. During this process, a template is created and stored either in a database on a smart card or any other storage medium. If the user is successfully enrolled, he or she is authorized to login using the biometrics. The new user template is converted by the dimensionality reduction module to a more compact representation in the feature vector space. The pattern is then given as an input to the associative memory which then used for neural network based matching. The outcome of the matching process is access granted or denied to the system (Yes/No). The dimensionality reduction module, the

Figure 1. The procedure for biometric system template generation from a training database

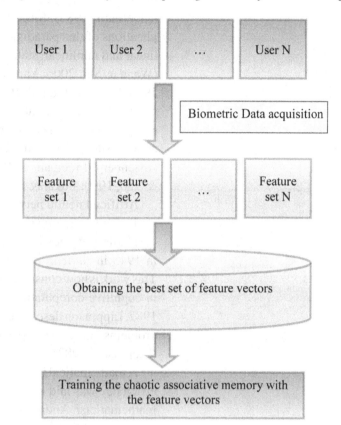

associative chaotic memory and the neural network operation are new intelligent methods in biometric research.

We will now describe *Artificial Neural Networks* (ANN) as one of new ways to improve biometric system security. An artificial neural network is typically defined as consisting of a large number of simple elements, called artificial neurons or nodes, which are linked by connections (Aihara, Takabe, & Toyoda, 1990). Together, all neurons cooperate to perform parallel distributed processing in order to solve a given computational problem. They are generally well suited for solving optimization tasks (Abu-Mostafa, 1986; Beck & Schlogl, 1995).

Neural networks provide a different paradigm in the area of intelligent computing. They are intended to simulate the structure and decision-making problem solving processes of a human brain, and thus their architecture is highly interconnected. However, this is far from being all their benefits. As with human brain, they are effective tools to make connections dynamic, i.e. changeable over time, and to place a different weighting to different elements of data stored in them. They can store various types of data, and thus allow high versatility in applications where samples might come from different sources. Moreover, the nature of neural networks makes them a perfect learner which allows identifying more prominent features and prevailing patterns in the collection of highly complex and frequently high-dimensional data.

All of those benefits make neural network a highly promising alternative to a traditional learner, especially in the context of biometric multi-modal system. Thus, in this chapter we focus on chaotic neural networks used for associative memory tasks in biometric applications. As it has

Figure 2. The procedure for user enrolment and matching

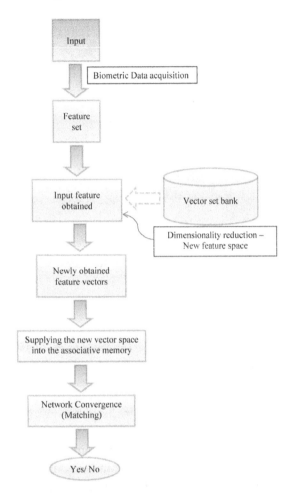

been recently shown in literature, these types of networks are proven to be very efficient tools in terms of time complexity and accuracy compared to traditional biometric systems (Ahmadian & Gavrilova, 2009, 2012a, 2012b).

3. NEURAL NETWORKS METHODOLOGY

3.1. Neural Network History

Artificial neural networks are computational models, where processing is performed by inter-related processing units. They were popularized

in the works by Hopfield (1990) and Choi and Hubernam (1983). Since then, research and applications of this paradigm became abundant for linear and nonlinear, static and dynamic systems (Wang & Shi, 2006; Yamada, Aihara, & Kotani, 1993; Yao, Freeman, Burke, & Yang, 1991). The nature of the network allows for a natural parallel computation and makes them an excellent learner as opposite to other traditional methodologies (Eisenberg, Freeman, & Burke, 1989; Freeman & Yao, 1990; Fukai & Shiino, 1990).

Artificial neural networks idea can be associated with the key work of McCulloch and Pitts (McCulloch & Pitts, 1943), which was published in 1943 in *Bulletin of Mathematical Biphysics*. This work is now considered to be a classical work in cognitive computational sciences domain. In 1987, Lippmann described how variety of applied problems can be solved with using neural nets (Lippmann, 1987).

A mathematical foundation for neural network efficiency was laid out by distinguished Russian mathematician Andrei Nikolaevich Kolmogorov, a graduate of the Moscow State University (Kolmogorov, 1957). His fundamental contribution was made available to neural network community by Spreecher (1993).

Another component without which the modern neural network landscape would be quite different was the paper titled "How brains make chaos in order to make sense of the world" (Skarda & Freeman, 1987). It links the chaos theory which became a foundation for chaotic neural networks (Chen & Aihara, 1995, 1997; Sandler & Yu, 1990; Wang & Smith, 1998) with neurological activities of the brain.

Since their development, neural networks has been found very useful in a number of applied problems, including pattern classification (Kittler, Hatef, Duin, & Matas, 1998; Lam & Suen, 1995), travelling salesman problem (Wang & Shi, 2006), face detection (Rowley, Baluja, & Kanade, 1998), large spatial databases (Ester, Kriegel, Sander, & Xu, 1996), clustering (Aihara, Takabe, & Toyoda,

1990), optimization and nonlinear system modelling (Abu-Mostafa, 1986).

3.2. Computational Effectiveness of Neural Networks

Neural networks possess some very nice properties making them an ideal choice for the complex problems, where computational complexity must be kept low (Lippmann, 1987). A quality of a solution to a typical computational problem might be estimated by a number of elementary operations, expressed as *time complexity*, required storage for both data and algorithm – *space complexity*, and computational resources needed to specify the algorithm – *Kolmogorov complexity*. In utilizing neural network for applied problems, the goal is to optimize the time, space, and Kolmogorov complexities of the neural network.

Thus, some examples of problems which can be solved by neural networks are pattern recognition problems in natural environments and artificial intelligence problems with large learning base. Another type of problem is biometric problem, which possesses all properties suitable for neural network approach: large datasets, complex optimization processing and pattern recognition.

4. CHAOS IN NEURAL NETWORKS

Chaos Online Dictionary definition is "complete disorder or confusion" or "Behavior so unpredictable as to appear random, owing to great sensitivity to small changes in conditions." In artificial neural network research, one of the first scientists to take a close look at chaos in biological patterns and to transpire this into computational domain was Freeman (Freeman & Yao, 1990; Yao, Freeman, Burke, & Yang, 1991).

Choi and Huberman reported that a system with both excitatory and inhibitory connections between neurons will display chaotic behavior (Choi & Huberman, 1983). The work on applying chaos in neural networks to avoid being trapped at a local minima and make them a superior computational method followed.

Nozawa demonstrated the search ability of chaotic neural network in (Nozawa, 1992). Chen and Aihara have introduced Chaotic Simulated Annealing method (CSA) (Chen & Aihara, 1995). Their method starts with a large negative self-coupling in the neurons and then gradually decreases it to stabilize the network. In 1990, Aihara and co-authors, have introduced a chaotic neural network which exhibits chaotic behavior using a negative self-coupling and gradually removing it in (Aihara, Takabe, & Toyoda, 1990). The model is based on the chaotic behavior of some biological neurons and has been successfully applied to several optimization problems such as TSP (Traveling Salesman Problem) (Yamada, Aihara, & Kotani, 1993) and BSP (Broadcast Scheduling Problem) (Wang & Shi, 2006). Chen and Aihara further introduced a chaotic simulated annealing approach, which was based on a transiently chaotic phase and a convergence phase, which can find the global minimums (Chen & Aihara, 1997). They also showed that the existence of attractors and network stability conditions suggested the dynamical phenomenon of crisis-induced intermittency to be the key principal for the chaotic switching among the minima problem.

This chapter creates a natural bridge from chaotic networks to recent biometric proposal research and utilities to the fullest advantages of chaotic neural networks.

5. FEATURE SPACE AND DIMENSIONALITY REDUCTION

Dimensionality reduction methods transform the data in the high-dimensional space to a space of fewer dimensions. The data transformation may be linear, as in principal components analysis, or non-linear. Many biometric spaces, such as facial biometrics, comprised of a large number of features which causes difficulties in learning and recognition process.

One of the main linear techniques for dimensionality reduction is a Principal Components Analysis (PCA), which performs a linear mapping of the data to a lower dimensional space in such a way, that the distinctiveness of the data in that low-dimensional space is maximized. However, the resulting dimensions might not be always efficient for biometric applications according to ambiguous subspaces problem. In this section, a survey on dimensionality reduction and methods of choosing proper feature subspaces has been provided. The argument to the use subspace-clustering dimensionality reduction is made based on extensive survey.

A basic solution to the general problem of clustering algorithms for high dimensional biometric data is to test all possible arbitrarily oriented subspaces for clusters. Obviously, there are an infinite number of arbitrarily oriented subspaces, so this naive solution is computationally infeasible. Rather, we need some heuristics and assumptions in order to conquer this infinite search space. A common approach to reduce the very large search space of all possible subspaces is to focus on axis-parallel subspaces only (Achtert, Bohm, Kriegel, Kroger, & Zimek, 2007). The authors argue that while an assumption that clusters can only be found in axis-parallel subspaces may be limiting, the advantage that the search space is restricted by the number of all possible axis-parallel subspaces makes it worth a try. The authors also provide an estimate on that number: in a *d*-dimensional dataset, the number of *k*-dimensional subspaces is

$$\binom{d}{k}(1 \le k \le d) \tag{9.1}$$

and the number of all possible subspaces is (Achtert, Bohm, Kriegel, Kroger, & Zimek, 2007):

$$\sum_{k=1}^{d}\binom{d}{k} = 2^d - 1 \tag{9.2}$$

According to (Achtert, Bohm, Kriegel, Kroger, & Zimek, 2007), there are four classes of clustering algorithms:

1. **Projected Clustering Algorithms:** These algorithms find a unique mapping of each point to exactly one subspace cluster and find the projection which clusters best the given set of points.
2. **Soft Projected Clustering Algorithms:** These algorithms assume that the number k of clusters is known in advance which allows to define objective function which results in an optimal set of k clusters.
3. **Subspace Clustering Algorithms:** These algorithms aim at locating all subspaces where clusters can be identified by localizing the search for relevant dimensions allowing to find clusters that exist in multiple, possibly overlapping subspaces.
4. **Hybrid Algorithms:** Finally, this forth class of algorithms tries to find some (not all) clusters that contain interesting subspaces instead of all subspace clusters.

In this chapter we focus on group 3 (Subspace clustering methods), which provide advanced tools for finding all subspace clusters of biometric data. They work well with the overlapping with overlapping subspaces and clusters and this is a very important fact for real biometric data analysis.

6. NEURAL-NETWORKS IN MULTI-MODAL BIOMETRICS

To validate the network and dimensionality reduction method, we have to describe to overall model of multi-biometric structure. A multi biometric system uses multiple sensors for data acquisition. This allows capturing multiple samples of a single biometric trait (called multi-sample biometrics) and/or samples of multiple biometric traits (called multi source or multi-modal biometrics). This also allows the system to enroll and authenticate a user who does not posses a specific biometric identifier.

6.1. Need for Dimensionality Reduction

One of the main problems faced by biometric community is the biometric system reliability and performance. Thus, the first goal is to find the principal components of the distribution of biometric features, or the eigenvectors of the covariance matrix of the set of biometric images. These eigenvectors can be thought of as a set of features which together characterize the variation between biometric samples. As the primary biometric samples, we select fingerprints and face images, since they provide significant variability in quality and a large number of multi-dimensional vectors.

Next, we can use subspace analysis and dimensionality reduction based on a generalized description of spherical coordinates. Each face image in the training set can be represented in terms of a linear combination of the eigenfaces. The number of possible eigenfaces is equal to the number of images in the training set. However, the biometric images can also be approximated using only the best eigenfaces—those that have the largest eigenvalues, and which therefore contribute to the most variance within the set of face images. The primary reason for using fewer eigenfaces is computational efficiency. This approach allows

not only move compact space representation, but also a convenient tool for subsequent clustering and learning of common patterns.

Next, we argue that neural networks are a fast and a reliable way for a biometric system to learn the pattern from the previously extracted subspaces. The neural network approach is based on the original chaotic noise injection strategy (Ahmadian & Gavrilova, 2009) which is the leading strategy for neural network training. The advantages are the ability to learn and later recognize new biometric samples in an unsupervised manner and that it is easy to implement using the proposed neural network architecture, as was shown in (Ahmadian & Gavrilova, 2012a, 2012b).

6.2. Overall System Architecture

The traditional architecture is shown in Figure 3.

One of the new system architectures that can be used for enhancing biometric system performance with neural network is shown in Figure 4.

A system implementation approach for multi-biometric recognition typically involves the following initialization operations:

1. Obtain a set of training images from users.
2. Calculate the eigenfaces from the training set, keeping the best M images with the highest eigenvalues. These N images define the biometric database.
3. Calculate the distribution in N-dimensional weight space for each training image by projecting those training images onto the biometric database.

Having initialized the system, the following steps are used to recognize the new biometric images:

1. Obtain the image to be recognized.

Figure 3. Multi-biometric security system architecture

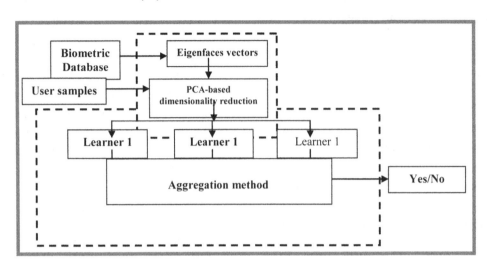

Figure 4. System architecture for biometric recognition system

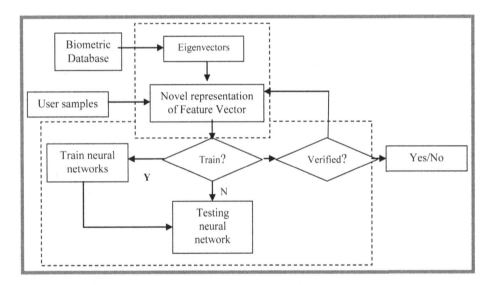

2. Calculate a set of weights of the *N* eigenfaces by projecting a new image onto each of the eigenfaces.

3. Compute the distance from the new image to other stored images in the biometric database in the vector space.

4. Classify the new image as either a known person or as unknown person.

5. Update the eigenfaces in the biometric database.

In this standard procedure, the key is the choice of the feature vector representing the image and also the distance to compute the closeness between the two feature vectors in the database. While the eigenvectors could be one of traditionally used features, in the multi-modal system the other parameters such as skin color, texture, characterizing features (glasses, beard, mustache), geometrical features (distance between the eyes, length of the nose, etc.), shape can be added to the equation. Each of those features can be given a weight based on how reliable this feature is or how distinctive it is. The distance between the vectors of features can be also computed differently. Each feature value can be converted to a binary representation: i.e. assigned values of 0 or 1. For instance, if distance between the eyes is below 15 cm – values of 0 is assigned, otherwise – value 1. Then the combined feature vector might have the following representation:

V1 (0,1,0,0,0,1,0,...., 1)

The second feature vector might be:

V2 (0,1,1,1,0,1,1,...., 1)

Their comparison can be done extremely fast by simply counting the number of the same bits (in the corresponding positions). It can be easily implemented in hardware.

Without resorting to binary representation, the distance between two high-dimensional vectors can be computed by computing the absolute values between two corresponding components. Note that the values do not necessarily have to be numerical. Such, the color of the eyes can be stored as blue, green, brown, for convenience.

In computing the difference between the values, either Euclidean norm can be used, or some other types of distances (Minkowski, geodesic, Mahalanobis, etc.). The choice of the distance metric is often dictated by such considerations as application domain, memory requirements, expectations on computational complexity, and ease of parallelization.

Finally, the vectors can be represented in a high-dimensional space and the transformation from that space onto one of other spaces can be done. It can be, as in the case with iris images, a polar coordinate transformation, or, as in the

case with clustering, projection of an image onto the sub-space identified by one of the methods described earlier (Section 5).

7. SUBSPACE ANALYSIS AND ASSOCIATE MEMORY

The input data to the multi-modal system is given as the set of images, corresponding to data dimensionality is then reduced by subspace analysis (Ahmadian & Gavrilova, 2012). The method for subspace analysis is based on a generalized description of spherical coordinates (Achtert, Bohm, Kriegel, Kroger, & Zimek, 2007). Generalized spherical coordinates combine $d - 1$ independent angles $á_1, ..., á_{d-1}$ with the norm r of a d-dimensional vector $x = (x_1, ..., x_d)^T$ to describe the vector x with respect to the given basis $e_1, ..., e_d$, where the x vector is the weight vectors obtained from the eigenfaces (Achtert, Bohm, David, Kroger, & Zimek, 2008).

Once the vector of weights from the candidate clusters is obtained, the next step is defining an energy model for an associative memory to learn the data patterns. The benefit is that this is a learner system that converges the given set of vectors to the stored pattern.

On-demand chaotic noise injection method (Ahmadian & Gavrilova, 2009) was developed to deal with a common drawback of non-autonomous methods - their *blind* noise-injecting strategy. The structure of noise function which is temporal function and is independent of the neighbor neurons, leads to the problems. In such methods, the chaotic or stochastic noises are injected into the network regardless of neuron's previous behavior.

In order to overcome the problem of blind noise injection in Chaotic Neural Network based models which may result in lower memory capacity for the huge amount biometric data, we use a new class of non-autonomous chaotic networks

(Ahmadian & Gavrilova, 2012). In such networks the key idea is that the chaotic noise injection is based on the behavior of neighbor neurons. This method is especially useful for a class of optimization problems in which state changing of neurons affects neuron with limited logical distances. In addition to feasibility of using such approach in biometric domain – both for single and multimodal biometric, the utilization of neural networks as a powerful learner was also demonstrated on example of both real human faces and artificial entities (Ahmadian & Gavrilova, 2012b).

8. PERFORMANCE OF NEURAL NETWORK ON FINGERPRINT MATCHING

The suggested methodology was successfully tested on fingerprint matching problem. The minutia extraction method which consists of estimation of orientation field, ridge detection, and minutia detection was applied, with geometrical aspects of the new method and the Hopfield Neural Network which is used for identification process.

Based on the original Delaunay triangulation method developed at Biometric Technologies laboratory (Wang, Luo, Gavrilova, & Rokne, 2007; Yanushkevich, Gavrilova, Wang, & Srihari, 2007), we perform fingerprint matching. The method relies on the structure of Delaunay triangles and uses both local and global criteria to match the minutiae. It also utilizes non-linear rigid function to simulate deformations of the fingerprint.

The new idea is based on the utilization of a Hopfield Neural Network (HNN) to retrieve patterns based on the previously stored fingerprints. However, there are a couple of issues to be considered. First is the fact that HNN is very sensitive to the misplacement of minutiae set and input noise, thus considering a Delaunay Triangulation (DT) as the pattern to be introduced to the network may

result in a high error rate. The second and the more important obstacle is the redundancy of the data introduced to the network by forcing it to learn the whole DT map of the minutiae set. In order to overcome these challenges, we have used the dual of the DT as the input pattern to the HNN.

By using duality, the method benefits from two the most important improvements. First, the input data has been used in the discretized space in order to distinct the location of resulted points from applying the duality method to the obtained DT. This sampling stage results in a more robust network, resistant to a significant amount of displacement and input error rate. The second improvement is the reduced rate of input data to the network. Specifically, instead of introducing a set of points as the discretized map to the network, we simply force the HNN to learn the equation of the lines for further retrievals.

Hopfield NNs are guaranteed to converge to a local minimum, but convergence to one of the stored patterns is not guaranteed. Hopfield nets can either have units that take on values of 1 or -1, or units that take on values of 1 or 0. So, the two possible definitions for unit i^{th} activation neuron are (Hopfield, 1990):

$$a_i \leftarrow \begin{cases} 1 & if \sum_j w_{ij}s_j > \theta_i \\ -1 & otherwise \end{cases}$$

$$a_i \leftarrow \begin{cases} 1 & if \sum_j w_{ij}s_j > \theta_i \\ 0 & otherwise \end{cases}$$

where, w_{ij} is the strength of the connection weight from unit j to unit i (also referred to as the weight of the connection), s_j is the state of unit j, θ_i is the threshold of unit i.

The connections in a Hopfield net have the following restrictions (Hopfield, 1990):

$w_{ij} = 0, \forall i$ (no unit has a connection with itself)

$w_{ij} = w_{ji}, \forall i, j$ (connections are symmetric)

Hopfield nets have energy E value associated with each state of the network (Hopfield, 1990):

$$E = -\frac{1}{2} \sum_{i<j} w_{ij} s_i s_j + \sum_i \theta_i s_i$$

According to the formula the network will converge to states which are local minima in the energy function. Training a Hopfield network means lowering the energy of the states that should be remembered. It is also called associative memory because the process is similar to remembering based on the similarity. This type of network can be easily modified to learn a visual pattern. Each neuron stands for one pixel in the image and all of the neurons are trained based on the input image, finally, the new pattern is introduced to the program and it will converge to the closest pattern in the memory.

The general goal is to train the network using the Delaunay triangulation of minutiae points. Since visual data is DT is not meaningful we took into account the concept of duality of change the line presentation of DT into a point presentation (See Figure 5).

A simple duality transform is the following:

Let $p \leftarrow (p_x, p_y)$ be a point in the plane. The dual of p, denoted p^*, is the line defined as

$$p^* \leftarrow (y = p_x x - p_y).$$

Note that in DT, one deals with line segments and not lines. The resulting pattern would be complicated to present, since the input pattern cannot be introduced to the Hopfield neural network because of the large number of colored pixels. For the sake of simplicity we assume each segment line in the corresponding DT as a line, so the resulting dual would be a single point. Some modifications in order to deal with perpendicular lines (very far corresponding duals) have been done and data have been normalized (See Figure 6).

Figure 5. Transforming a DT to its dual, the shape on the right is used for training the Hopfield neural network

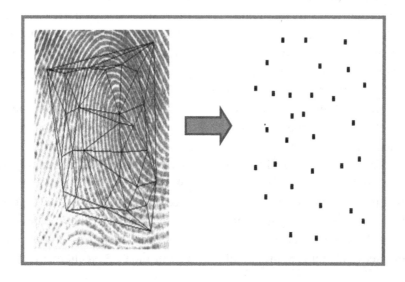

Figure 6. Interface showing the Hopfield neural network-based matching process

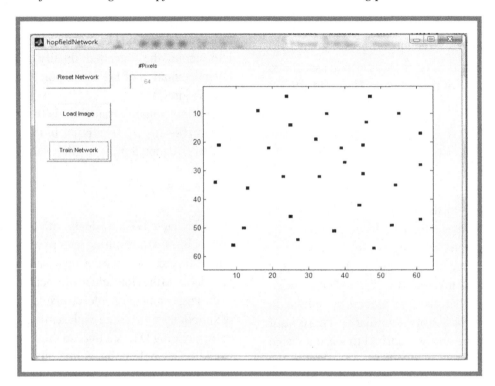

9. CNN-BASED MINUTIAE MATCHING METHOD

We now provide the experimental results to showcase advantages of chaotic neural network for fingerprint recognition in both accuracy and circumvention (resistance to errors). The test database contains both high and very low quality fingerprint images, thus if high level of accuracy is confirmed on those samples, the system possesses both precision and circumvention. The first task was to find the minutiae set based on the algorithm given in previous section. A binarized version of the input image is created next. The finding process is done based in the ridge with one pixel width, so the in the next process the thinning filter will be performed on the image.

After calculating the minutiae set and implementing the Delaunay Triangulation algorithm for this set, the dual of Delaunay Triangulation is found and it is presented as the input of the neural network. Afterwards, the trained network finds the closest stored pattern to the provided pattern. The experiments were performed on the database from the Biometric Systems at the University of Bologna, which consists of 21*8 fingerprint images of 256*256 sizes.

We set the size of the neural network to 64*64, the reduces size would act as a sampling method, while the Hopfield neural network is a minimization function for the energy function which is based on the spatial distance of the introduced in saved patterns.

The performance of the proposed method was shown to be superior in comparison with traditional method such as DT matching, DT matching with ridge geometry and standard minutiae method (Wang, Luo, Gavrilova, & Rokne, 2007). For the provided samples the false rejection rate is much smaller (which means much better) than of previous methods—it show improvement from one of the best recent algorithms based on rigid

Delaunay triangulation from around 5% to 0.047%. The false acceptance rate remains very low, at 1%. The experiments showed that utilizing neural network improves the FRR rate significantly and can retain FAR at a very low rate.

10. SUMMARY

The chapter introduces the concept of neural network as an intelligent learning tool for biometric security system.

It starts off by reviewing the history of neural nets and the most common applications of this method in a variety of computational and optimization problems. It then proceeds to the basic overview of chaos in neural network. It points out on-demand chaotic noise injection method was developed to deal with a common drawback of non-autonomous methods—their *blind* noise-injecting strategy.

The second part of the chapter presents the issue of high-dimensionality in the context of a complex biometric security system. The amount of data and its complexity can be overwhelming, and as one way of dealing with that the dimensionality reduction techniques can be used. They are typically based on clustering or transformations from one space to another.

The reduced dimensionality vector can be then used in the energy model for an associative memory which will learn the data patterns. The benefit is that this is a learner system that converges the given set of vectors to the stored pattern in a network, which can be later used for biometric recognition and also for identifying the most significant biometric patterns.

The chapter concludes with some examples where in addition to feasibility of using such approach in biometric domain - both for single and multi-modal biometric. It is worth noticing that the utilization of neural networks as a powerful learner can also be extended to artificial entities.

REFERENCES

Abu-Mostafa, Y. S. (1986). Neural networks for computing? In J. S. Denker (Ed.), *Neural Networks for Computing, 151*, 1-6.

Achtert, E., Bohm, C., David, J., Kroger, O., & Zimek, A. (2008). Robust clustering in arbitrarily oriented subspaces. In *Proceedings of 8th SIAM International Conference on Data Mining*, (pp. 763-774). SIAM.

Achtert, E., Böhm, C., Kriegel, H. P., Kröger, P., & Zimek, A. (2007). On exploring complex relationships of correlation clusters. In *Proceedings of the 19th International Conference on Scientific and Statistical Database Management*, (pp. 7-21). IEEE.

Ahmadian, K., & Gavrilova, M. (2012b). A multi-modal approach for high-dimensional feature recognition. *The Visual Computer*. doi: doi:10.1007/s00371-012-0741-9.

Ahmadian, K., & Gavrilova, M. L. (2009). On-demand chaotic neural network for broadcast scheduling problem. []. ICCSA.]. *Proceedings of the ICCSA, 2*, 664–676.

Ahmadian, K., & Gavrilova, M. L. (2012a). Dealing with biometric multi-dimensionality through chaotic neural network methodology. *International Journal of Information Technology and Management, 11*(1/2), 18–34. doi:10.1504/IJITM.2012.044061.

Aihara, K., Takabe, T., & Toyoda, M. (1990). Chaotic neural networks. *Physics Letters. [Part A], 144*(6-7), 333–340. doi:10.1016/0375-9601(90)90136-C.

Beck, C., & Schlogl, F. (1995). *Thermodynamics of chaotic systems*. Cambridge, UK: Cambridge University Press.

Chakrabarti, K., Keogh, E. J., Mehrotra, S., & Pazzani, M. J. (2002). Locally adaptive dimensionality reduction for indexing large time series databases. *ACM Transactions on Database Systems*, 27(2), 188–228. doi:10.1145/568518.568520.

Charu, C. Aggarwal, & Philip, S. Y. (2008). A framework for clustering uncertain data stream. In *Proceedings of the 24th International Conference on Data Engineering*, (pp. 150-159). IEEE.

Chen, L., & Aihara, K. (1995). Chaotic simulated annealing by a neural network model with transient chaos. *Neural Networks*, 8(6), 915–930. doi:10.1016/0893-6080(95)00033-V.

Chen, L., & Aihara, K. (1997). Chaos and asymptotical stability in discrete time neural networks. *Physica D. Nonlinear Phenomena*, 104, 286–325. doi:10.1016/S0167-2789(96)00302-8.

Choi, M. Y., & Huberman, B. A. (1983). Dynamic behavior of nonlinear networks. *Physical Review A.*, 28, 1204–1206. doi:10.1103/PhysRevA.28.1204.

Daleno1, D., Dellisanti, M., & Giannini, M. (2008). Retinal fundus biometric analysis for personal identifications. In *Proceedings of ICIC*, (pp. 1229-1237). Springer.

Eisenberg, J., Freeman, W. J., & Burke, B. (1989). Hardware architecture of a neural network model simulating pattern recognition by the olfactory bulb. *Neural Networks*, 2, 315–325. doi:10.1016/0893-6080(89)90040-3.

Ester, M., Kriegel, H. P., Sander, J., & Xu, X. (1996). A density-based algorithm for discovering clusters in large spatial databases with noise. In *Proceedings of 2nd International Conference on Knowledge Discovery*, (pp. 226-231). IEEE.

Freeman, W. J., & Yao, Y. (1990). Model of biological pattern recognition with spatially chaotic dynamics. *Neural Networks*, 3, 153–170. doi:10.1016/0893-6080(90)90086-Z.

Fukai, T., & Shiino, M. (1990). Asymmetric neural networks incorporating the dale hypothesis and noise-driven chaos. *Physical Review Letters*, 64, 1465–1468. doi:10.1103/PhysRevLett.64.1465 PMID:10041402.

Han, J., & Kamber, M. (2001). *Data mining concepts and techniques*. San Francisco, CA: Kaufmann.

Ho, T. K., Hull, J. J., & Srihari, S. N. (1994). Decision combination in multiple classifier systems. *IEEE Transactions on Pattern Analysis and Machine Intelligence*, 16(1), 66–75. doi:10.1109/34.273716.

Hopfield, J. J. (1990). The effectiveness of analogue neural network hardware. *Network (Bristol, England)*, 1(1), 27–40. doi:10.1088/0954-898X/1/1/003.

Jain, A. K., Ross, A., & Pankanti, S. (1999). A prototype hand geometry-based verification system. In *Proceedings of International Conference on Audio- and Video-based Biometric Person Authentication*, (pp. 166-171). IEEE.

Jain, A. K., Ross, A., & Prabhakar, A. (2004). An introduction to biometric recognition. *IEEE Transactions on Circuits and Systems for Video Technology*, 14(1), 4–20. doi:10.1109/TCSVT.2003.818349.

Johnson, R. G. (1991). *Can iris patterns be used to identify people?* Los Alamos, CA: Chemical and Laser Sciences Division Los Alamos National Laboratory.

Kittler, J., Hatef, M., Duin, R. P., & Matas, J. G. (1998). On combining classifiers. *IEEE Transactions on Pattern Analysis and Machine Intelligence, 20*(3), 226–239. doi:10.1109/34.667881.

Kolmogorov, A. N. (1957). On the representation of continuous functions of several variables by superposition of continuous functions of one variable and addition. *Doklady Akademii: Nauk USSR, 114,* 679–681.

Lam, L., & Suen, C. Y. (1995). Optimal combination of pattern classifiers. *Pattern Recognition Letters, 16,* 945–954. doi:10.1016/0167-8655(95)00050-Q.

Lam, L., & Suen, C. Y. (1997). Application of majority voting to pattern recognition: An analysis of its behavior and performance. *IEEE Transactions on Systems, Man, and Cybernetics. Part A, Systems and Humans, 27*(5), 553–568. doi:10.1109/3468.618255.

Lee, M. W., & Ranganath, S. (2003). Pose-invariant face recognition using a 3D deformable model. *Pattern Recognition, 36*(8), 1835–1846. doi:10.1016/S0031-3203(03)00008-6.

Lippmann, R. P. (1987). An introduction to computing with neural nets. *IEEE Magazine on Accoustics, Signal, and Speech Processing, 4,* 4–22.

McCulloch, J. L., & Pitts, W. (1943). A logical calculus of ideas immanent in nervous activity. *The Bulletin of Mathematical Biophysics, 5,* 115–133. doi:10.1007/BF02478259.

Moon, Y. S., Yeung, H. W., Chan, K. C., & Chan, S. O. (2004). Template synthesis and image mosaicking for fingerprint registration: An experimental study. In *Proceedings of International Conference on Acoustic Speech and Signal Processing,* (vol. 5, pp. 409–412). IEEE.

Nozawa, H. (1992). A neural-network model as a globally coupled map and applications based on chaos. *Chaos (Woodbury, N.Y.), 2*(3), 377–386. doi:10.1063/1.165880 PMID:12779987.

Rowley, H. A., Baluja, S., & Kanade, T. (1998). Neural network-based face detection. *IEEE Transactions on Pattern Analysis and Machine Intelligence, 20*(1), 23–38. doi:10.1109/34.655647.

Sandler, & Yu, M. (1990). Model of neural networks with selective memorization and chaotic behavior. *Physics Letters A, 144,* 462-466.

Skarda, C. A., & Freeman, W. J. (1987). How brains make chaos in order to make sense of the world. *The Behavioral and Brain Sciences, 10,* 161–195. doi:10.1017/S0140525X00047336.

Spreecher, D. A. M. (1993). A universal mapping for Kolmogorov's superposition theorem. *Neural Networks, 6*(8), 1089–1094. doi:10.1016/S0893-6080(09)80020-8.

Verlinde, P., & Cholet, G. (1999). Comparing decision fusion paradigms using k-NN based classifiers, decision trees and logistic regression in a multi-modal identity verification application. In *Proceedings of Second International Conference on Audio- and Video-Based Biometric Person Authentication (AVBPA),* (pp. 188–193). AVBPA.

Wang, C., Luo, Y., Gavrilova, M. L., & Rokne, J. (2007). Fingerprint image matching using a hierarchical approach. In Nedjah, N., Abraham, A., & de Macedo Mourelle, L. (Eds.), *Computational Intelligence in Information Assurance and Security* (pp. 175–198). Berlin, Germany: Springer. doi:10.1007/978-3-540-71078-3_7.

Wang, L., & Shi, H. (2006). A gradual noisy chaotic neural network for solving the broadcast scheduling problem in packet radio networks. *IEEE Transactions on Neural Networks, 17*(4), 989–1001. doi:10.1109/TNN.2006.875976 PMID:16856661.

Wang, L., & Smith, K. (1998). On chaotic simulated annealing. *IEEE Transactions on Neural Networks*, *9*, 716–718. doi:10.1109/72.701185 PMID:18252495.

Yamada, T., Aihara, K., & Kotani, M. (1993). Chaotic neural networks and the travelling salesman problem. In *Proceedings of International Joint Conference on Neural Networks*, (pp. 1549–1552). IEEE.

Yanushkevich, S., Gavrilova, M., Wang, P., & Srihari, S. (2007). *Image pattern recognition: Synthesis and analysis in biometrics*. New York, NY: World Scientific Publishers.

Yao, Y., Freeman, W. J., Burke, B., & Yang, Q. (1991). Pattern recognition by a distributed neural network: An industrial application. *Neural Networks*, *4*, 103–121. doi:10.1016/0893-6080(91)90036-5.

Chapter 10
Novel Applications of Multimodal Biometrics

ABSTRACT

This chapter presents the original idea of using social networks and context information in multimodal biometric for increased system security. A recently investigated study's outcomes is presented, which showcase this idea as a new step in multi-biometric research. Since this method does not degrade the performance of the system and is not computationally expensive, it can be used in any biometric framework. However, as the amount of improvement depends on how distinctive and predictable people are in terms of their behavioral patterns, the method is most suitable for the predictable environments with some predefined behavioral routines. Fine tuning the system for each environment to find the most suitable parameters based on the behavioral patterns of that specific environment can result in better performance. This research is validated on example of gait recognition.

1. INTRODUCTION

The idea of using social network for multimodal biometric only recently has made its way into state-of-the-art multimodal biometric research. In the previous sections, we proposed to use chaotic neural network for better trait learning, and presented the dimensionality reduction technique for simplifying biometric template to ease the burden

on computational resources. We further suggested to combine biometric data with information about people including "soft biometric." This section introduces one more idea—idea of using social connections—to the mix.

Let us consider this idea for a moment. The person identity can be determined not only through passport data (eye color, height, weight, birth date, nationality, etc.), and not solemnly from finger-

DOI: 10.4018/978-1-4666-3646-0.ch010

prints or iris, but also through associations with others. This, to the common questions on identity estimates: "What person knows," "What person possesses," and "Who person is" which correspond to Knowledge, ID Possession and Biometrics, we add a novel component:

"Whom Person Knows"

Abundance of social networks and sites, used for communication, news and information exchange (Facebook, MySpace, Classmates, LinkedID, Twitter, MSNLive, etc.) has created a highly favorable environment for connecting people regardless of their physical appearance, geographical location, age, religion, job etc., based on some commonly shared interests or trends. These are those trends that we are specifically interested in assisting to determine what person identity is.

Let us examine the principles under which Social networks works. Normally, individual users need to create their profiles containing certain bibliographical information about themselves. To protect user privacy, social networks usually have controls that allow users to choose who can view their profile, who may contact them, who can add them to their list of contacts, etc. Users can upload images to their profiles, post blog entries for others to read, search for other users with similar interests, share lists of contacts, follow threads of discussions, celebrate birthdays, give virtual cards or flowers etc. These social connections, friends, groups of interests, contacts, discussion threads or event favorite TV shows can, in combination, identify individual as good if not better than usual password/ID/biometric does. The difficulty in exploiting this type of information as primary cue is, however, abundant. The difficulties are:

- Social data needs to be gathered from social sites, and filtered to render it suitable for subsequent identification processes.
- Variety of social networks and different types of data/connections in networks need to be identified and represented uniformly.

- Certain social traits need to be chosen over others to render better recognition results.
- Social features need to be studies and standards on them developed.
- Social networks and on-line communities are highly volatile phenomena which might be available today but not tomorrow due to server/network connection, administrator/host, maintenance/migration, number of users.
- High computational power is needed for identification.

However, to counteract those negative trends, some features of social networks make them ideal candidates as *supplementary traits for multi-biometrics*.

They include:

- Unique set of interests, which is highly valuable for person identification.
- Unique network of close friends for each individual which can be used for recognition.
- Data is abundant.
- Data is accessible, usually freely shared by network users.
- It can be collected/processed remotely at any moment of time, and does not requires any specialized expensive hardware to collect.

Moreover, secondary uses of such information can shine light not only on security or person identification, but also on other scientific research (ethics, consumer surveys, psychology, learning, collaborative environments, virtual reality, art).

2. GAIT ANALYSIS IN MULTI-BIOMETRIC RESEARCH

Let us now introduce one more biometric—based on Human Gait—and explain how its performance can be augmented using social traits.

Gait analysis deals with analyzing the patterns of walking movement. The fundamental work in gait analysis is attributed to Johansson (1973) who showed that people can quickly recognize the motion of walking only from the moving patterns of a few point lights attached to a human body. Stimulated by Johansson's work, Cutting and Kozlowski in (Cutting & Kozlowski, 1977) did some experiments to show that the same array of point lights can be used to recognize friends even if they happen to have similar height, width and body shapes. Considering the wide variety of potential applications for gait analysis, these studies open the gate for a lot of research in the field. Although gait analysis is most well-known for its application in access control, surveillance and activity monitoring, it can also be used in sports training to analyze the athlete's movements and give suggestions for improvement. Medical sciences can also take advantage of gait analysis techniques in diagnosing and maybe even developing some strategies to treat patients with walking disorders.

The focus of gait analysis is the same as for any other biometric: identifying people from the way they walk. Gait recognition has recently attracted more attention due to a set of unique and interesting properties. This trait is unobtrusive meaning that the attention or cooperation of the subject is not needed for collecting the data (Wang, 2005). Unlike many other biometrics, no specially designed hardware is needed. A surveillance camera is sufficient for data acquisition. Data collection can be both overt or covert, i.e. it takes place with or without knowledge of the subject. Next, this trait is remotely observable and the subject does not even need to be close to the camera (Wang, 2005). In addition, imitating the walking style of another person is quite difficult. It is also not always possible to conceal the way a person walks (Liu & Zheng, 2007). Finally, gait recognition techniques usually do not need high resolution video sequences (Wang, She, Nahavandi, & Kouzani, 2010) and, since they usually

work on binary silhouettes, they are not extremely sensitive to illumination changes (Cuntoor, Kale, & Chellappa, 2003).

However, as any biometric, gait recognition suffers from limitations and challenges. Age, mood, illness, fatigue, drug, or alcohol consumption can affect the walking style of a person (Liu & Zheng, 2007). Person's walking style might be also affected by conditions, such as type of shows, pavement, etc (Bashir, Xiang, & Gong, 2009). In summary, the main drawback of using gait for individual identification compared to other biometrics is its wide variability per subject.

The main idea thus is to augment the imperfect biometric such as Gait recognition by using supplementary metadata about people's social life/social connections.

3. LITERATURE SURVEY

Gait recognition is gaining momentum in biometric research. It is remotely observable, hard to imitate and hard to conceal trait, which has a number of unique characteristics. Although a lot of research has been dedicated to gait recognition in the last few decades, most of current gait recognition system still work under very constrained conditions and their performance in real scenarios is not perfect.

To deal with environmental changing conditions, one common approach in biometric is to *improve* algorithms by making them more robust to appearance changes, lighting conditions, noise, etc. This approach, however, can result in extremely complex algorithms. An alternative approach is to use a common gait recognition algorithm and combine it with *social information* about people. For this purpose, students at Biometric Technology Laboratory, University of Calgary have recently implemented a gait recognition system based on Gait Energy Images (Bazazian & Gavrilova, 2012). The context of the video such as location (indoor, outdoor), carrying conditions (suitcase, coffee, backpack, etc.) and time of the day (morning, afternoon, evening, and night) has been used as supplementing metadata to improve gait recognition accuracy. Different ways of modeling the social context and behavioral trends of each individual have been investigated. The output of the gait recognition system has been combined with the video context to make the final decision. Performance evaluation of the system shows that using behavioral patterns of people's social life always improves the accuracy of the gait recognition system. The amount of improvement depends on how discriminative the social patterns of the people are. If the people under study have very predictable and distinctive behavioral patterns, it's even possible to achieve 100% recognition rate.

4. DETAILED METHODOLOGY

A gait recognition system typically includes the following parts:

1. Subject detection and silhouette extraction.
2. Gait cycle detection.
3. Feature extraction.
4. Feature selection and/or dimensionality reduction.
5. Recognition.

In the following, each of these building blocks will be described.

4.1. Subject Detection and Silhouette Extraction

The first step of gait recognition is to detect the subject in the image and separate it from the background. The most popular technique for this purpose is background subtraction. The first step of the background subtraction method is to learn a background model. This model represents the background color for each pixel. The background model can be learned beforehand. It can be as simple as one single image usually obtained as the mean or median of all the frames or, in more complicated scenarios, it can include color distributions such as Gaussian distributions for each pixel. For making the model robust to lighting changes, it is possible to make it dynamic by updating it on a frame by frame basis. Once the model is built, any pixel with the distance from the background model greater than a certain threshold is considered as a foreground pixel. Some post processing might also follow afterwards to remove noise. This post processing usually includes erosion, dilation, and finding the largest connected component of the foreground image.

4.2. Gait Cycle Detection

Gait can be treated as a signal. To make the gait recognition algorithm robust to speed changes, it is necessary to obtain the gait period. The features are usually extracted for one single gait cycle. Therefore, before extracting the features it is essential to find the starting and the ending frame of the gait cycle or, in other words, to partition the video sequence into gait cycles. Many cycle detection algorithms are based on counting the number of foreground pixels in certain regions. By detecting two subsequent minimum or maximum, it is often possible to find the gait cycles (Sarkar, Phillips, Liu, Vega, Grother, & Bowyer, 2005).

4.3. Feature Extraction

After detecting the subject and partitioning the sequence into gait cycles, the next step is to extract some useful features. There are generally two main approaches for feature extraction: model-based and model-free.

The *model-based approaches* use an explicit model to represent the human body (Wang, She, Nahavandi, & Kouzani, 2010). These methods estimate the parameters of the model in each frame. The value of these parameters and how they change over time is used for gait representation.

These methods have a couple of advantages. First, model-based methods are view and scale invariant (Wang, She, Nahavandi, & Kouzani, 2010). Instead of using the silhouettes directly, these methods fit a model to the silhouette. Consequently, the size of the silhouette and its viewing direction in case of a 3D model does not have much influence on the recognition output. They also can to some degree deal with occlusion and self-occlusion. Since body parts are modeled separately, even if some of the parts are not visible due to occlusion there is still a chance that other parts are visible.

Therefore, the algorithm would not lose the subject and the parameters of the visible body parts can be used to estimate the parameters of the other invisible ones. Finally, for the same reasons, model-based methods are not extremely sensitive to appearance changes, such as carrying condition. Having a priori knowledge about how the human body should look like, these methods are able to detect if a person is carrying an object and then exclude that object from their calculations. However, the model-based approaches have the number of disadvantages. First, due to highly flexible structure of non-rigid human body and also self occlusion (Yang, Zhou, Zhang, Shu, & Yang, 2008), the search space is huge and estimating the model's parameters is tremendously difficult (Wang, Zhang, Pu, Yuan, & Wang, 2010). As a result, these methods are generally computationally expensive and extremely time consuming (Wang, She, Nahavandi, & Kouzani, 2010). Second, since the estimation of the model parameters needs high quality videos, these methods are usually sensitive to the quality of the video sequences and vulnerable to noise (Wang, She, Nahavandi, & Kouzani, 2010). A number of methods in this category has been developed (Mishra & Erza, 2010; Yoo & Nixon, 2011; Ma, Wang, Nie, & Qiu, 2007). An example of walking process and corresponding stick figure is found (Figure 1 (Han & Bhanu, 2006)).

4.4. Model-Free Approaches

Model free approaches do not use a priori body model. Instead, they make a compact representation for the walking motion by considering the silhouette as a whole. The model free approaches have a number of advantages. First, these methods are cheap and fast (Wang, She, Nahavandi, & Kouzani, 2010). Unlike model-based methods, they do not need to estimate the parameters of a model in each frame and the processing needed to be done in each frame is generally negligible compared to model based approaches. Similarly not needing to search for the parameters of a complicated model in a huge search space, these methods usually only need some general information about the silhouette shape and they are not sensitive to noise and the quality of the video sequences (Wang, She, Nahavandi, & Kouzani, 2010). Furthermore, since these methods usually work only on silhouettes, no other information (color, texture, grayscale values, etc.) is needed for their processing. As a result they can be used for gait recognition at night using infrared imagery (Liu & Zheng, 2007).

However, model free approaches also suffer from a couple of disadvantages. Since these methods do not have any a priori knowledge about the human body and they only work based on the silhouette shape, they are generally more sensitive to factors that can change the appearance of the person and his silhouette. This includes loose clothing, wearing a hat, carrying an object, etc. (Ma, Wang, Nie, & Qiu, 2007). For the exact same reasons, these methods are also sensitive to view and scale changes (Wang, She, Nahavandi, & Kouzani, 2010).

There are generally two main categories of model free approaches (Wang, Zhang, Pu, Yuan, & Wang, 2010):

Figure 1. Normalized and centralized binary silhouettes and their corresponding gait energy (Han & Bhanu, 2006)

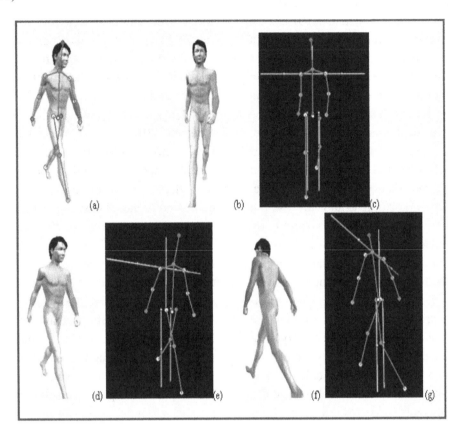

Temporal Sequence

These methods extract features from each frame and the sequence of the extracted features is used as the final feature vector. In other words, these methods represent gait as a temporal sequence. These methods need a lot of storage space to store the resulting feature sequences for each subject. Furthermore, they also need to train a sequence matching algorithm (like HMM) for each subject. This algorithm will be used to recognize if an input temporal sequence matches that person's walking style. Training such a framework usually needs a lot of training data and is also time consuming. Additionally, these methods need complex sequence matching for final recognition which can be both computationally expensive and also time consuming (Wang, Zhang, Pu, Yuan, & Wang, 2010).

Single Template

These methods, similarly, extract the features from each frame but in the end all the features are combined together into one template. In other words, they represent the gait cycle by one single template. By doing so, they save storage space and also computation time (Han & Bhanu, 2006). Therefore, they create a very compact representation (Liu & Zheng, 2007). The resulting template should satisfy the following properties (Boulgouris & Chi, 2007): capturing structural information, capturing dynamic information, provide a compact representation with small number of features, be robust to speed changes.

The single template matching methods are not extremely sensitive to silhouette noise, holes, shadows, and missing parts (Liu & Zheng, 2007; Bashir, Xiang, & Gong, 2009). But they are sensitive to appearance changes (Bashir, Xiang, & Gong, 2009). Knowing both advantages and disadvantages of model free and model based approaches, the fact that model free methods are cheap, fast and easy to understand has made them more popular and in fact most of gait recognition algorithms fall in this category. Sharma et al. in (Sharma, Tiwari, Shukla, & Singh, 2011) simply use the whole silhouette sequence as their feature vector. Since this method saves the whole silhouette sequence for each person, it needs a lot of storage space.

Han and Bhanu introduced the idea of Gait Energy Image (GEI) in (Han & Bhanu, 2006). GEI is the average of all silhouette images for a single gait cycle. An example of this approach is shown in Figure 2. To avoid over fitting and to make the method more robust to little distortions including shadows, missing body parts, scale changes, etc. they generate some synthetic gait energy images by adding distortion to the lower part of the real gait energy images for each individual. GEI is an efficient and compact representation of gait and it also reduces the noise by averaging (Yang, Zhou, Zhang, Shu, & Yang, 2008) but it does not have any temporal information about the motion. Liu and Zheng in (Liu & Zheng, 2007) developed the idea of GEI and proposed Gait History Images (GHIs) to compensate this problem. The idea comes from Motion History Images (MHIs) (Bobick & Davis, 2001). They calculate motion images by obtaining the difference of each two subsequent frames and then obtain MHI as a weighted combination of all the motion images. The weight of each motion image comes from its position in the gait cycle. Therefore, the parts that have moved recently appear brighter in the

resulting MHI. These motion history images as shown in Figure 2 only represent the moving parts of the human body however in the gait analysis problem the static parts of the human body and its shape is also useful for recognition. Therefore, they find the static parts of human body by calculating the intersection of all silhouettes through the gait cycle and add them to the MHI to obtain their Gait History Image (GHI). The difference between these different temporal templates is depicted in Figure 3. Motion Energy Image (MEI) is obtained by finding the intersection of all the motion images. MEI represents which pixels have been moved during the walking motion (Bobick & Davis, 2001).

F´elez et al. in (Martin-Felez, Mollineda, & Sanchez, 2011), also in an attempt to improve the performance of GEI, segment the gait cycle into four key poses. They extract the silhouette in each frame and classify that frame as one of the four key poses. Subsequently, they obtain a Gait Energy Image (GEI) for each key pose by calculating the average of all the frames assigned to that key pose. Each GEI is classified separately using a nearest neighbor classifier and the final person identification is done by majority voting of the decisions of different classifiers.

Chen et al. in (Chen, Liang, & Zhao, 2009) propose a new approach for gait recognition with the purpose of reducing the effect of silhouette incompleteness and distortion. They cluster the gait cycle by simply putting adjacent frames in the same cluster. The number of clusters is determined based on the amount of distortion. For each cluster, the GEI is calculated and then denoised by removing the pixels with values less than a threshold to obtain DEI image for that cluster. Afterwards, for each frame the frame difference is calculated by subtracting two subsequent frames and the representation of that frame is obtained by adding the corresponding DEI image to the positive portion of the frame difference image. They use frieze and Haar wavelet coefficients of the resulting sequence as their features. Liu and Sarkar

Figure 2. From left to right: motion energy image, motion history image, gait energy image, and gait history image (Liu & Zheng, 2007)

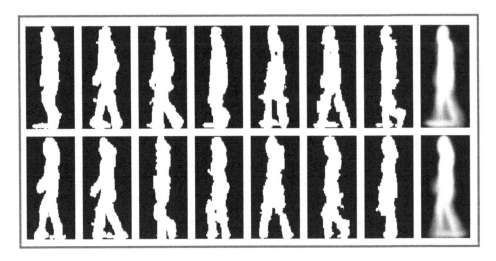

Figure 3. Number of pixels against the number of frames

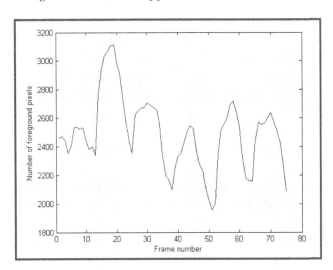

in (Liu & Sarkar, 2004) extract the silhouettes using background subtraction and then align the silhouettes and calculate their average over one gait cycle to obtain their final template. The resulting template mainly captures information about the shape of the body. The temporal dynamic aspect of the gait is also captured to some extent. Due to the high popularity of Gait Energy Images (Han & Bhanu, 2006) and in general image averaging techniques in gait feature extraction, they did some experiments to investigate the importance

of different parts of silhouette in the final average template and showed that the lower part of the silhouette which contains leg dynamics is as important as the upper part of the silhouette which mainly captures the shape of the body.

In a similar work, Veres et al. (Veres, Gordon, Carter, & Nixon, 2004) investigated the importance of different parts of silhouette for gait recognition. According to their experiments for the methods that use silhouette averaging including Gait Energy images, the upper part of the silhouette (head and body), which mostly represents the static part of the gait, is the most important information for gait recognition.

Gait Energy Images have also been augmented with soft biometrics. Moustakas et al. (Moustakas & Starvopoulos, 2010) combine soft biometric features with gait features to improve the recognition rate by reducing the search space. The used soft biometric features are the subject's height and the stride length. Gait Energy Image and Radon transforms are used as geometric gait features. These features are combined using probabilistic approach. Boulgouris and Chi in (Boulgouris & Chi, 2007) calculate the Radon transform of each silhouette image for different directions. They combine the resulting Radon transforms to obtain a final template for each gait cycle.

4.5. Feature Selection

The features extracted from the silhouettes are usually high-dimensional. Working with huge feature vectors, comparing them and storing them is computationally expensive, time consuming and needs a lot of storage space. Consequently, almost always a dimensionality reduction method is used to find most dominant features and remove redundant or less important ones. Some of the typical popular dimensionality reduction techniques for gait recognition are PCA and its variants (such as MDA) (Sharma, Tiwari, Shukla, & Singh, 2011):

Principal Component Analysis (cPCA)

PCA is a projection that transforms the data to a lower dimensional space by using the data correlations. This transformation is usually obtained by Eigenvalue analysis of the data covariance matrix.

Multiple Discriminant Analysis (MDA)

While PCA looks for the best subset of variables for representing the data, MDA finds the optimal transformation for separating the data by maximizing the ratio of the between class distances to the within class distances. This approach is also based on Eigenvalue analysis and is employed in our work.

In some applications, a combination of these approaches is used to achieve the best compression possible. Other common methods for multidimensional data reduction include clustering and transformation to other coordinates. In clustering, while k-mean is being one of the popular methods, the Voronoi diagram based clustering is an alternative based on topological space representation (Yuan, Gavrilova, & Wang, 2008), as well as subspace clustering described in previous Chapter for biometric data dimensionality reduction.

4.6. Recognition

The method used for final recognition is generally dependent on the feature extraction algorithm. When the output of feature selection method produces a temporal sequence, state space models are usually used for finding the similarity of temporal sequences. Hidden Markov Model (HMM) is the popular model used for this purpose. However, if the output of the feature selection phase is one single template, then finding the distances between the templates can be used for recognition. Nearest-neighbor classifiers and support vector machines are two of most popular methods in this category (Wang, She, Nahavandi, & Kouzani, 2010).

Final recognition in Sharma, Tiwari, Shukla, and Singh (2011) is done by calculating the Euclidean distance of silhouette sequences. Han and Bhanu (2006) obtain the synthetic and real GEIs of the probe, calculate the Euclidean distances with the corresponding GEI templates and fuse the results. Liu and Zheng (2007) also use Euclidean distance between Motion History Images for final recognition. Félez et al. (Martin-Felez, Mollineda, & Sanchez, 2011) use a set of nearest neighbor classifiers and the final person identification is done by majority voting of the decisions of different classifiers. Nearest-neighbor classifier has also been used (Xu & Zhang, 2010; Chen, Huang, Guo, & Dong, 2010; Ma, Wang, Nie, & Qiu, 2007; Lam & Lee, 2006; Yoo & Nixon, 2011) for person identification.

Frieze and Haar wavelet coefficients as features has been used (Chen, Liang, & Zhao, 2009). Decision fusion for gait recognition was utilized (Cuntoor, Kale, & Chellappa, 2003). The idea behind decision fusion is to extract different features, match them separately and then combine the results of different matchers to make the final decision. The features used in this work were (Cuntoor, Kale, & Chellappa, 2003):

- Left and right projections of the silhouette to capture the motion of hands and legs.
- Width vector of front view sequences to capture changes in the subject's height.
- Width vector of the lower part of the silhouette to capture leg dynamics.

The final decision was made by combining the resulting similarity values using sum, product and min operators.

4.7. Performance Evaluation

For evaluating the performance of the system, the dataset is divided into two subsets: one for training and one for testing. Training set or gallery is a set that is used for training the system. This set represents the gait samples of subjects known to the system. The test set, however, includes some unknown subjects that will be presented to the system later for recognition. The output of the system is usually a ranked list of the people best matched with the presented unknown subject. Frequently, two performance measures are used to report the performance of the system:

- **Rank1 Performance:** The percentage of the times that correct subject appeared as the first subject of the rank list.

- **Rank5 Performance:** The percentage of the times that the correct subject appears among the top 5 ranks.

5. SOCIAL NETWORKS FOR MULTIBIOMETRIC RESEARCH

Social Network refers to a group of individual entities and connections among them. In formal representation of social network as a graph, individuals are called nodes (vertices) and connections between them are edges (representing contacts). The goal of Social Network Analysis (SNA) is to measure relationships between connected knowledge entities (individuals, groups, organizations). (Moreno, 1934).

SNA can successfully handle multiple relations and connections. Additional benefits are derived by marking certain edges/connections or assigning weights, thus allowing to model either the strength of the relationship or the type of relationship (i.e. friend/colleague). Numerous social networks are available such as LinkedIn (network of professionals and colleagues), Facebook (network of friends/family), Twitter (group of people with similar interests and followers), etc.

Social network can be commonly represented as a graph $G = \{(vi, vj)\}$, where G refers to the name of the network and (vi, vj) is the pair of vertices in the network. An adjacency matrix representation is often used to represent such a graph.

The graph theory is highly useful in social networks analysis. It helps to focus on important components of the network, understand connections and identify important nodes, cluster data, realize patterns, and compare different networks. Thus, network paths can be explored, including maximum connectivity paths, all connected paths, longest/shortest paths etc. A more important direction in which multi-biometric research is mostly comes useful is the relationship and distance between biometric features. The goal of biometric research is to identify trends that are unique and best characterize given instance. In social network, relationship among nodes becomes important and thus the importance of the node itself can be computed and is referred to as centrality (Freeman, 1979).

There are some centrality measures identified in literature (Koschutzki, Lehmann, Peeters, Richter, Tenfelde-Podehl, & Zlotowski, 2005). The *degree centrality* of node *k* is a measure of the number of contacts that node *k* has. In a case of multibiometric and social networks, a node with a high degree centrality represents a highly connected, highly social, or highly important individual. The common types of centrality measures are (Koschutzki, Lehmann, Peeters, Richter, Tenfelde-Podehl, & Zlotowski, 2005):

- **The Relative Centrality:** Which represents an importance of the node.
- **The Betweenness Centrality:** Which is expressed as a probability that the shortest path between any pair of nodes would go through the given node.
- **The Closeness Centrality:** Which represents connectivity factor of the node also dependent on the type of distance measure.
- **The Eigenvector Centrality:** Which is based on the principle that connections to high-influence nodes contribute more to the overall importance of the node. As reported by Wikipedia, Google's PageRank is a variant of the Eigenvector centrality measure.

Utilizing social networks can be beneficial for biometric research. Benefits of fusing social context with one biometric – Gait, is showcased in the next section.

6. FUSING SOCIAL CONTEXT WITH GAIT RECOGNITION

Using one of the approaches discussed in the previous section, we can get information on location, time, and carrying condition of all persons in the video sequences in the database, including both training and test datasets. Gait recognition method employed here is based on *Gait Cycle detection*. For extracting the gait cycle, in each frame we count the number of pixels in the lower half of the silhouettes (leg region). Using these values, we obtain a curve.

An example of this curve is shown in Figure 3. It is easy to see that the minimums of this curve correspond to the points that the legs are together or the beginning of a cycle. Therefore, all the frames between two minima, skipping every other minimum, belong to the same cycle. Using this property, one can find the local minima of this curve to identify the beginning and ending frames of the cycle. We can also compute the gait cycle as the average of distances between minima, skipping every other minimum. The model-free approach based on Gait Energy Images is used for feature extraction and PCA for feature dimension reduction.

The method is combined with contextual data. We can find out which times, locations and carrying conditions are possible for each person and use this extra information to improve the performance of our gait recognition system. As mentioned in the previous section, the output of the gait recognition system is a rank list of top five candidates. For each candidate, its rank and its matching score (similarity of its feature to the probe's feature calculated using rank level methodology) are known. We use the context information to reorder the candidates in the rank list. For each candidate, using the virtual context data available for that person we examine if the time, location and carrying condition of the probe (the unknown video presented to the system) are possible for that person. Based on the context data, for each parameter we assign a score to the candidate if that parameter value is possible for that person. We then add all the scores for all three parameters to assign a context score to each candidate. Having the matching score and context score, we first normalize the scores to the range [0..1] and then simply add them to find the final score of each candidate. We then use the final scores to reorder the candidates in the rank list.

The information can be both extracted from row data (video sequences) or stored in supplementary social traits database. Additional information such as appearance of both subjects who are connected in Social Network together within short period of time in the same location can be a hint to increase recognition rate. Degree of Social Network connectivity (such as Eigenvector centrality or closeness centrality) is also taken into account through assignment of weights. In addition to using context data from video sequences or subject database, direct link to on-line search engine can be established as to have most up to date profile of the subject. This however is direction for the future research.

For evaluating the system, half of the dataset is used as the training set and the other half as the test dataset. We train the system using the training dataset and then evaluate the resulting system using both training and test datasets. The performance of the system is reported in terms of rank 1 performance and rank 5 performance, as discussed in the performance evaluation section of the literature review. We first calculate the rank 1 performance and rank 5 performance of the gait recognition system without using the context data. Afterwards, we run the system one more time and this time we combine the context data with the output of the gait recognition system as discussed in the previous section. We again compute the rank 1 performance and rank 5 performance and compare their values to what we had before to

investigate the impact of using the context data on system's performance.

The overall architecture of the proposed multi-biometric system for gait recognition based on fusing context data of the subjects with gait recognition algorithm is shown in Figure 4.

The module responsible for extracting gait features in the form of GEI and the flow of information corresponding to its operations is visualized in Figure 5.

7. IMPLEMENTATION DETAILS AND RESULTS

The system has been implemented in MATLAB. The gait dataset used for evaluating the system is Dataset A of CASIA gait database. This dataset is one of the most popular datasets available for gait recognition. Dataset A contains 20 subjects (persons) and 12 different sequences per each subject captured at different viewing angles. This dataset is publicly available online. Since the Gait Energy Images mainly work for side-view video sequences, we only use the side-view sequences of each person. Therefore, for each person have only 4 video sequences. Half of this data is used for training and the rest for testing. Consequently, the training set consists of all 20 subjects and 2 video sequences per subject. Meaning, all the individuals are known to the system. Similarly, the test set also consists of 20 subjects and 2 video sequences per subject. The performance of the system is evaluated using both the training and testing sets.

The result of the system performance reported in terms of rank 1 performance and rank 5 performance as discussed in the performance evaluation section.

In the first case the Gait Energy images are directly used as the features and in the second case, PCA is applied on the Gait Energy Images and the resulting 20 PCA coefficients are used as the feature vectors. Based on Table 1 using PCA

Figure 4. The flowchart of the context-based gait recognition system

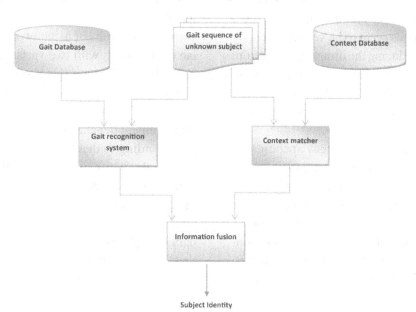

Figure 5. The flowchart of the gait recognition module

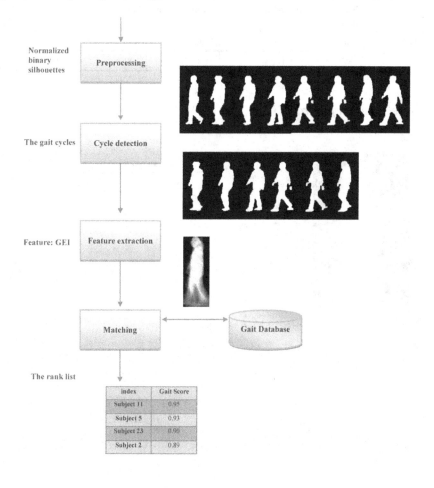

adversely affects the performance of the system. Having only 20 classes can be one of the main reasons. The classes seem to be very different by themselves and reducing the dimensionality from 6400 to 20 is maybe more than enough to lose. Based on these results, we decided to use directly Gait Energy Images as our features in the rest of experiments.

When subjects do not have any specific patterns of behavior and act randomly, recognition performance is not improved. However, it is not degraded either, thus making system applicable in real-life security applications. If subjects follow certain patterns in their workflow, carrying conditions, schedule, or connection to other subjects, their patterns of behavior can be modeled using

Gaussian distributions, and we can observe tremendous improvements in the recognition accuracy. The results of using two different Gaussian distributions with two variances is shown in Table 1. For the case that the variance value is one, it was even possible to achieve 100% recognition rate. These results show that the amount of improvement depends on how distinctive people are according to their behavioral patterns.

It is also worth mentioning that the amount of computation this method introduces to the existing gait recognition system is very negligible. In fact, we only need to take a look at the context data of 5 people and add some scores based on that. Therefore, this can be done in constant time. Furthermore, if we use profiles for modeling the

Table 1. The results of adding context data (random context database and Gaussian context database)

No context database		Random virtual context database		Gaussian (variance=2)		Gaussian (variance=1)	
Rank 1	Rank 5	Rank 1	Rank 5	Rank 1	Rank 5	Rank 1	Rank 5
80%	100%	83%	100%	88%	100%	100%	100%

Figure 6. The gait recognition program GUI

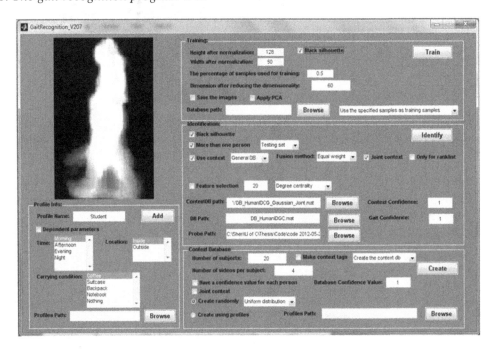

behavioral patterns, even creating the context database is not that time consuming and computationally expensive.

The program user interface with the set of various options is shown in Figure 6.

8. SUMMARY

The section presents an original idea of using social networks and context information in multimodal biometric for increased system security. Preliminary results recently obtained showcase that this is a new step in multi-biometric research. In terms of the possible applications of the proposed method, since this method does not degrade the performance of the system and is not computationally expensive, it can be used anywhere a gait recognition system can be used. However, as the amount of improvement depends on how distinctive and predictable people are in terms of their behavioral patterns, the method is most suitable for the predictable environments with some predefined behavioral routines. Fine tuning the system for each environment to find the most suitable parameters based on the behavioral patterns of that specific environment can result in better performance.

REFERENCES

Bashir, K., Xiang, T., & Gong, S. (2009). Gait recognition using gait entropy image. In *Proceedings of the 2009 IEEE International Conference on Crime Detection and Prevention*. IEEE.

Bazazian, S., & Gavrilova, M. (2012). Context-based gait recognition. In *Proceedings of SPIE*. SPIE.

Bobick, A. F., & Davis, J. W. (2001). The recognition of human movement using temporal templates. *IEEE Transactions on Pattern Analysis and Machine Intelligence, 23*(3), 257–267. doi:10.1109/34.910878.

Boulgouris, N. V., & Chi, Z. X. (2007). Gait recognition using radon transform and linear discriminant analysis. *IEEE Transactions on Image Processing, 16*(3), 731–740. doi:10.1109/TIP.2007.891157 PMID:17357733.

Chen, C., Liang, J., & Zhao, H. (2009). Frame difference energy image for gait recognition with incomplete silhouettes. *Pattern Recognition Letters, 30*, 977–984. doi:10.1016/j.patrec.2009.04.012.

Chen, S., Huang, W., Guo, Q., & Dong, L. (2010). Wavelet moments for gait recognition represented by motion templates. In *Proceedings of International Conference on Fuzzy Systems and Knowledge Discovery*, (pp. 620-624). IEEE.

Cuntoor, N., Kale, A., & Chellappa, R. (2003). Combining multiple evidences for gait recognition. In *Proceedings of the 2003 International Conference on Multimedia and Expo*, (pp. 113-116). IEEE Computer Society.

Cutting, J. K., & Kozlowski, L. K. (1977). Recognizing friends by their walk: Gait perception without familiarity cues. *Bulletin of the Psychonomic Society, 9*(5), 353–356.

Freeman, L. C. (1979). Centrality in social networks: Conceptual clarification. *Social Networks, 1*, 215–239. doi:10.1016/0378-8733(78)90021-7.

Han, J., & Bhanu, B. (2006). Individual recognition using gait energy image. *IEEE Transactions on Pattern Analysis and Machine Intelligence, 28*(2), 316–322. doi:10.1109/TPAMI.2006.38 PMID:16468626.

Johansson, G. (1973). Visual perception of biological motion and a model for its analysis. *Perception & Psychophysics*, *14*(2), 201–211. doi:10.3758/BF03212378.

Koschutzki, D., & Lehmann, K. A. Peeters, L., Richter, S., Tenfelde-Podehl, D., & Zlotowski, O. (2005). Centrality indices. In Proceedings of Network Analysis, (pp. 16–61). Berlin, Germany: Springer.

Lam, T. H. W., & Lee, R. S. T. (2006). A new representation for human gait recognition: Motion silhouettes image (MSI). In *Proceedings of the International Conference on Biometrics*, (pp. 612–618). IEEE.

Liu, J., & Zheng, N. (2007). Gait history image: A novel temporal template for gait recognition. In *Proceedings of the 2007 IEEE International Conference on Multimedia and Expo*, (pp. 663-666). Beijing, China: IEEE.

Liu, Z., & Sarkar, S. (2004). Simplest representation yet for gait recognition: averaged silhouette. In *Proceedings of the 17th International Conference on Pattern Recognition*, (pp. 211-214). IEEE.

Ma, Q., Wang, S., Nie, D., & Qiu, J. (2007). Recognizing humans based on gait moment image. In *Proceedings of the Eighth ACIS International Conference on Software Engineering, Artificial Intelligence, Networking, and Parallel/Distributed Computing*, (pp. 606-610). IEEE Computer Society.

Martín-Félez, R., Mollineda, R. A., & Sánchez, J. S. (2011). Human recognition based on gait poses. *Lecture Notes in Computer Science*, *6669*, 347–354. doi:10.1007/978-3-642-21257-4_43.

Mishra, P., & Erza, S. (2010). Human gait recognition using Bezier curves. *International Journal on Computer Science and Engineering*, *3*(2), 969–975.

Moreno, J. L. (1934). *Who shall survive?* Washington, DC: Nervous and Mental Disease Publishing Company.

Moustakas, K. T., & Stavropoulos, D. G. (2010). Gait recognition using geometric features and soft biometrics. *IEEE Signal Processing Letters*, *17*(4), 367–370. doi:10.1109/LSP.2010.2040927.

Sarkar, S., Phillips, P. J., Liu, Z., Vega, I. R., Grother, P., & Bowyer, K. W. (2005). The HumanID gait challenge problem: Data sets, performance, and analysis. *IEEE Transactions on Pattern Analysis and Machine Intelligence*, *27*(2), 162–177. doi:10.1109/TPAMI.2005.39 PMID:15688555.

Sharma, S., Tiwari, R., Shukla, A., & Singh, V. (2011). Frontal view gait based recognition using PCA. In *Proceedings of the International Conference on Advances in Computing and Artificial Intelligence*, (pp. 124-127). ACM.

Veres, G. V., Gordon, L., Carter, J. N., & Nixon, M. S. (2004). What image information is important in silhouette-based gait recognition? In *Proceedings of the 2004 IEEE Conference on Computer Vision and Pattern Recognition*, (pp. 776-782). IEEE.

Wang, C., Zhang, J., Pu, J., Yuan, X., & Wang, L. (2010). Chrono-gait image: A novel temporal template for gait recognition. In *Proceedings 11th European Conference on Computer Vision*, (pp. 257-270). IEEE.

Wang, C.-H. (2005). *A literature survey on human gait recognition techniques. Directed Studies EE8601*. Toronto, Canada: Ryerson University.

Wang, J., She, M., Nahavandi, S., & Kouzani, A. (2010). A review of vision-based gait recognition methods for human identification. In *Proceedings of the 2010 Digital Image Computing: Techniques and Application*, (pp. 320-327). Piscataway, NJ: IEEE.

Xu, S.-L., & Zhang, Q.-J. (2010). Gait recognition using fuzzy principal component analysis. In *Proceedings of the Second IEEE International Conference on e-Business and Information System Security*. IEEE.

Yang, X., Zhou, Y., Zhang, T., Shu, G., & Yang, J. (2008). Gait recognition based on dynamic region analysis. *Signal Processing, 88*(9), 2350–2356. doi:10.1016/j.sigpro.2008.03.006.

Yoo, J.-H., & Nixon, M. (2011). Automated markerless analysis of human gait motion for recognition and classification. *ETRI Journal, 33*(2), 259–266. doi:10.4218/etrij.11.1510.0068.

Yuan, L., Gavrilova, M., & Wang, P. (2008). Facial metamorphosis using geometrical methods for biometric applications. *International Journal of Pattern Recognition and Artificial Intelligence, 22*(3), 555–584. doi:10.1142/S0218001408006399.

Chapter 11
Conclusion

ABSTRACT

This chapter summarizes the book's contribution to the domain of information security and biometric research, and points out future directions of this dynamic field of studies. The book is the first to link the concepts of security, biometric and computational intelligence, and show how intricately they are woven with one another.

1. BOOK SUMMARY

This book is the first to link the concepts of security, biometric and computational intelligence and show how intricately they are woven with one another. It starts off by looking at the history of computational intelligence and relates it to current security research. Biometric systems, and their functionalities, performances issues and challenges are reviewed next. Based on known issues with single biometrics, a concept of multimodal biometric system is presented. Pros and cons of multimodal biometric systems along with various development issues have been studied in literature. They are further discussed and illustrated in this book.

Among the fusion approaches, pre-matching fusion approaches, such as sensor level fusion and feature level fusion, are discussed. Match score level fusion methods which are very popular with developers are presented and decision level fusion approaches used in many commercial biometric systems are discussed next. The focus then moves onto rank level fusion methods. Novel rank fusion approaches based on fuzzy logic and Markov Chain method have been introduced. For rank level fusion, comparison with highest rank, Borda count, logistic regression methods have been performed.

Outcomes of the practical multi-modal system implementation and experimentations have been presented and discussed further. Aside from increased accuracy, enrolment and response times

DOI: 10.4018/978-1-4666-3646-0.ch011

which are very essential in time critical security systems have been computed and analyzed for different scenarios.

Novel alternative approaches based on computational intelligence paradigm are presented next. These include chaotic neural network and dimension-reduction concepts for multi-biometric system design, robotic biometric and avatar recognition for intelligent software security systems, and a concept of application of soft biometric, social networks and social trends for improved performance of multi-modal biometric system.

2. CONCLUSION

The world where we are living is more complex and yet arguably less secure than ever. There is tremendous pressure on government and public organizations to increase the security, with increased investments to both research and development in this field. However, at the same time, the security breaches become more frequent, more costly and more severe, with potential to negatively impact all spheres of personal life of an individual, a large corporations, stock markets, and even political structure of a country.

Multimodal biometric system emerged as one of the solutions which provides a significant advantages over unimodal biometric system, such as increased and reliable recognition performance, fewer enrolment problems and suitability of the final system to be used in real world security crucial applications. The design of a multimodal biometric system is a challenging task due to heterogeneity of the biometric sources in terms of the type of information, the magnitude of information content, correlation among the different sources and conflicting performance requirements of the practical applications.

Researchers are still trying to find a good combination of biometric traits and various fusion methods to find the optimum recognition performance. As recent results has shown, a multimodal biometric system developed using fuzzy fusion approaches along with soft biometric information is a powerful tool to achieve this. A novel rank fusion approach, Markov chain based rank fusion, has been introduced which satisfies the Condorcet criteria essential for any fair rank aggregation process and can enhance the performance by a significant margin even in the presence of some low quality data. It significantly improves the response time as well as the recognition performance of any multimodal biometric system.

Furthermore, utilizing neural networks and social traits can lead to increased system reliability and recognition rates. A combination of such intelligent approaches with formal topological neighborhood relationship and rigorous mathematical measures of closeness seem to result in highly reliable systems.

3. FUTURE RESEARCH DIRECTIONS

There are many interesting questions left for future research enquiry. Many of them were identified in concluding chapter discussions. Some key observations are summarized here.

A true multimodal database can be quite useful for developing a reliable and efficient security application. True multimodal database with the identical conditions can be employed for further performance analysis. In most cases, biometric based security systems need to operate in real-time mode. Proper instruments for capturing real-time data and peripheral communications are needed for this purpose. Special concentration is needed to automatically acquire soft biometric information in real-time setup.

Dual or tri-level (different fusion in different levels of the system) fusion scenarios can be implemented to make the system faster and significantly reduce the error rate. More advanced rules for fuzzy multimodal system can further improve performance rates. Combining multi-biometric with social traits is promising new direction of research.

There are many opportunities to investigate other approaches to intelligent computing, including clustering, dimensionality-reduction, neural networks, evolutionary intelligence, patter matching, and other learning methods in the presence of complex biometric traits. Moving towards distributed systems based on enquiry-based principles will further benefit the developer while open new opportunities for system development.

In context-based social network analysis, an efficient scheme which can successfully and uniquely model behavior of an individual in a distinctive way is still hard to come by. Investment of time and resources into this challenging problem seems to be a sensible thing to do.

There is an exponential growth of data which can be collected about individual or business in a modern society from a variety of sources (demo-graphic data, financial rating, web browsing history, club memberships, frequent buyer programs, social and professional networks). This calls for a higher attention to individual rights on privacy and security, on both legislative and academic fronts. Such areas as combining encryption with data protections protocols, and investing into biometric template protection or biometric cancellability, become research areas of the future.

Applying biometric principals to artificial entities has emerged as a new domain with its own set of problems and unique challenges. Finally, new applications in other areas, such as on-line security, gaming, virtual worlds and intelligent software systems and robots provide rich and exciting domain for future scientific research and enquiries.

Appendix 1: List of Figures

Appendix 2: List of Tables

Appendix 3: List of Abbreviations

AI: Artificial Intelligence
ANN: Artificial Neural Network
ASM: Active Shape Model
ATT: Automated Turing Test
BSP: Broadcast Scheduling Problem
BTLab: Biometric Technologies Laboratory
CAPTCHA: Completely Automated Public Turing Test to tell Computers and Humans Apart
CI: Computational Intelligence
CMC: Cumulative Match Characteristics
CNN: Chaotic Neural Network
CSA: Chaotic Simulated Annealing
CT: Computed Tomography
DNA: Deoxyribonucleic Acid
DT: Delaunay Triangulation
EER: Equal Error Rate
FAR: False Accept(ion) Rate
FMM: Fuzzy Membership Map
FRR: False Reject(ion) Rate
FTCR: Failure-to-Capture Rate
FTER: Failure-to-Enroll Rate
GA: Genetic Algorithm
GAR: Genuine Accept Rate
GDA: Generalized Discriminant Analysis
GEI: Gait Energy Image
GHI: Gait History Image
GRR: Genuine Reject Rate
HCI: Human-Computer Interface
HIP: Human Interactive Proof
HMM: Hidden Markov Model
HNN: Hopfield Neural Network
ICA: Independent Component Analysis
KDDA: Kernel Direct Discriminant Analysis
LDA: Linear Discriminant Analysis

LE: Laplacian Eigenmaps
MVU: Maximum Variance Unfolding
MC: Markov Chain
MDA: Multiple Discriminant Analysis
MHI: Motion History Image
MHP: Mandatory Human Participation
MMP: Machine Maintenance Problem
MOEO: Multi-Objective Evolving Object
MRI: Magnetic Resonance Imaging
MS: Multidimensional Scaling
NBC: Naïve Bayes Classifier
NN: Neural Network
PCA: Principal Component Analysis
PNG: Portable Network Graphics
PSO: Particle Swarm Optimization
RBF: Radial Basis Function
RBFNN: Radial Basis Function Neural Networks
ROC: Receiver Operating Characteristics
RTT: Reversed Turing Test
SI: Swarm intelligence
SNA: Social Network Analysis
SS: Subspace Clustering
SVM: Support Vector Machine
TSP: Traveling Salesman Problem
VD: Voronoi Diagram

Compilation of References

360. Biometrics. (2012). *Website*. Retrieved online from http://360biometrics.com/faq/Keystroke_Keyboard_Dynamics.php

Aarabi, P., & Dasarathy, B. V. (2004). Robust speech processing using multi-sensor multi-source information fusion: An overview of the state of the art. *Information Fusion*, *5*, 77–80. doi:10.1016/j.inffus.2004.02.001.

Abaza, A., & Ross, A. (2009). Quality based rank-level fusion in multibiometric systems. In *Proceedings of 3rd IEEE International Conference on Biometrics: Theory, Applications and Systems*. Washington, DC: IEEE.

Abu-Mostafa, Y. S. (1986). Neural networks for computing? In J. S. Denker (Ed.), Neural Networks for Computing, 151, 1-6.

Achtert, E., Bohm, C., David, J., Kroger, O., & Zimek, A. (2008). Robust clustering in arbitrarily oriented subspaces. In *Proceedings of 8th SIAM International Conference on Data Mining*, (pp. 763-774). SIAM.

Achtert, E., Böhm, C., Kriegel, H. P., Kröger, P., & Zimek, A. (2007). On exploring complex relationships of correlation clusters. In *Proceedings of the 19th International Conference on Scientific and Statistical Database Management*, (pp. 7-21). IEEE.

Ahmadian, K., & Gavrilova, M. (2012). A multi-modal approach for high-dimensional feature recognition. *The Visual Computer*. doi: doi:10.1007/s00371-012-0741-9.

Ahmadian, K., & Gavrilova, M. L. (2009). On-demand chaotic neural network for broadcast scheduling problem. []. ICCSA.]. *Proceedings of the ICCSA*, *2*, 664–676.

Ahmadian, K., & Gavrilova, M. L. (2012). Dealing with biometric multi-dimensionality through chaotic neural network methodology. *International Journal of Information Technology and Management*, *11*(1/2), 18–34. doi:10.1504/IJITM.2012.044061.

Ahn, L. V., Blum, M., Hopper, N., & Langford, J. (2003). CAPTCHA: Using hard AI problems for security. In Proceedings of Eurocrypt, 2003. Eurocrypt.

Ahn, L. V., Blum, M., Hopper, N., & Langford, J. (2003). CAPTCHA: Using hard AI problems for security. *Lecture Notes in Computer Science*, *2656*, 294–311. doi:10.1007/3-540-39200-9_18.

Aihara, K., Takabe, T., & Toyoda, M. (1990). Chaotic neural networks. *Physics Letters. [Part A]*, *144*(6-7), 333–340. doi:10.1016/0375-9601(90)90136-C.

Ailon, N., Charikar, M., & Newman, A. (2005). Aggregating inconsistent information: Ranking and clustering. In *Proceedings of 37th Annual ACM Symposium on Theory of Computing (STOC)*, (pp. 684-693). Baltimore, MD: ACM.

Ailon, N. (2007). Aggregation of partial rankings, p-ratings and top-m lists. *Algorithmica*, *57*(2), 284–300. doi:10.1007/s00453-008-9211-1.

Apu, R. A., & Gavrilova, M. L. (2012). Battle swarm: The genetic evolution of tactical strategies and battle efficient formations. *ACM Transactions on Autonomous and Adaptive Systems*. Retrieved from http://3ia.teiath.gr/3ia_previous_conferences_cds/2006/Papers/Full/Apu24.pdf

Apu, R., & Gavrilova, M. (2006). Battle swarm: An evolutionary approach to complex swarm intelligence. In *Proceedings of the 9th International Conference on Computer Graphics and Artificial Intelligence*, (pp. 139-150). Limoges, France: Eurographics.

Asoh, H., Hayamizu, S., Hara, I., Motomura, Y., Akaho, S., & Matsui, T. (1997). Socially embedded learning of the office-conversant mobile robotjijo-2. In *Proceedings of 15th International Joint Conference on Artificial Intelligence*. ACM/IEEE.

Bailly-Baillire, E., Bengio, S., Bimbot, F., Hamouz, M., Kittler, J., & Marithoz, J. Thiran, J. P. (2003). The BANCA database and evaluation protocol. In *Proceedings of International Conference on Audio- and Video-Based Biometric Person Authentication*, (pp. 625-638). Guildford, UK: IEEE.

Baird, H. S., & Bentley, J. L. (2005). Implicit CAPTCHAs. In Proceedings *of the SPIE/IS&T Conference on Document Recognition and Retrieval XII (DR&R2005)*. San Jose, CA: SPIE/IS&T.

Bartlett, M. S., Movellan, J. R., & Sejnowski, T. J. (2002). Face recognition by independent component analysis. *IEEE Transactions on Neural Networks, 13*(6), 1450–1464. doi:10.1109/TNN.2002.804287 PMID:18244540.

Bashir, K., Xiang, T., & Gong, S. (2009). Gait recognition using gait entropy image. In *Proceedings of the 2009 IEEE International Conference on Crime Detection and Prevention*. IEEE.

Baudat, G., & Anouar, F. (2002). Generalized discriminant analysis using a kernel approach. *Neural Computation, 12*(10), 2385–2404. doi:10.1162/089976600300014980 PMID:11032039.

Bazazian, S., & Gavrilova, M. (2012). Context-based gait recognition. In *Proceedings of SPIE*. SPIE.

Bebis, G., Deaconu, T., & Georiopoulous, M. (1999). Fingerprint identification using Delaunay triangulation. In Proceedings of ICIIS99, (pp. 452-459). ICIIS.

Beck, C., & Schlogl, F. (1995). *Thermodynamics of chaotic systems*. Cambridge, UK: Cambridge University Press.

Belhumeur, P., Hespanha, J., & Kriegman, D. (1997). Eigenfaces vs. fisherfaces: Recognition using class specific linear projection. *IEEE Transactions on Pattern Analysis and Machine Intelligence, 19*(7), 711–720. doi:10.1109/34.598228.

Benitez, A. B., & Chang, S. F. (2002). Multimedia knowledge integration, summarization and evaluation. In *Proceedings of Workshop on Multimedia Data Mining*, (vol. 2326). Springer.

Bentley, J., & Mallows, C. L. (2006). CAPTCHA challenge strings: Problems and improvements. In *Proceedings of Document Recognition & Retrieval*. IEEE.

Biometric News Portal. (2012). *Website*. Retrieved from http://www.biometricnewsportal.com/biometrics_benefits.asp

Biometrics. (2009). *Emerging devices technical brief*. New York, NY: AT&T.

Black, D. (1963). *The theory of committees and elections* (2nd ed.). Cambridge, UK: Cambridge University Press.

Bobick, A. F., & Davis, J. W. (2001). The recognition of human movement using temporal templates. *IEEE Transactions on Pattern Analysis and Machine Intelligence, 23*(3), 257–267. doi:10.1109/34.910878.

Bolle, R. M., Connell, J. H., Pankanti, S., Ratha, N. K., & Senior, A. W. (2004). *Guide to biometrics*. New York, NY: Springer-Verlag.

Bonabeau, E., Dorigo, M., & Theraulaz, G. (1999). *Swarm intelligence: From natural to artificial systems*. Oxford, UK: Oxford University Press.

Borda, J. C. (1781). *M'emoire sur les 'elections au scrutin*. Paris, France: Histoire de l'Acad'emie Royale des Sciences.

Boulgouris, N. V., & Chi, Z. X. (2007). Gait recognition using radon transform and linear discriminant analysis. *IEEE Transactions on Image Processing, 16*(3), 731–740. doi:10.1109/TIP.2007.891157 PMID:17357733.

Boyd, R. S. (2010). *Feds thinking outside the box to plug intelligence gaps*. Retrieved from http://www.mcclatchydc.com/2010/03/29/91280/feds-thinking-outside-the-box.html

Burge, M., & Burger, W. (1998). Ear biometrics. In Jain, A. K., Bolle, R., & Pankanti, S. (Eds.), *Biometrics: Personal Identification in Networked Society* (pp. 273–286). Norwell, MA: Kluwer Academic Publishers.

Cappelli, R., Maio, D., & Maltoni, D. (2001). Modelling plastic distortion in fingerprint images. *Lecture Notes in Computer Science*, *2013*, 369–376. doi:10.1007/3-540-44732-6_38.

Chaabane, L., & Abdelouahab, M. (2011). Improvement of brain tissue segmentation using information fusion approach. *International Journal of Advanced Computer Science and Applications*, *2*(6), 84–90.

Chakrabarti, K., Keogh, E. J., Mehrotra, S., & Pazzani, M. J. (2002). Locally adaptive dimensionality reduction for indexing large time series databases. *ACM Transactions on Database Systems*, *27*(2), 188–228. doi:10.1145/568518.568520.

Chang, K., Bowyer, K., & Barnabas, V. (2003). Comparison and combination of ear and face images in appearance-based biometrics. *IEEE Transactions on Pattern Analysis and Machine Intelligence*, *25*, 1160–1165. doi:10.1109/TPAMI.2003.1227990.

Charles, J. S., Rosenberg, C., & Thrun, S. (1999). Spontaneous, short-term interaction with mobile robots. In *Proceedings of IEEE International Conference on Robotics and Automation*, (pp. 658-663). IEEE.

Charu, C. Aggarwal, & Philip, S. Y. (2008). A framework for clustering uncertain data stream. In *Proceedings of the 24th International Conference on Data Engineering*, (pp. 150-159). IEEE.

Chechik, G., & Tishby, N. (2003). Extracting relevant structures with side information. *Advances in Neural Information Processing Systems*, *15*.

Chen, K.-J., & Barthes, J.-P. (2008). Giving an office assistant agent a memory mechanism. In *Proceedings of 7th IEEE International Conference on Cognitive Informatics*, (pp. 402 – 410). IEEE.

Chen, S., Huang, W., Guo, Q., & Dong, L. (2010). Wavelet moments for gait recognition represented by motion templates. In *Proceedings of International Conference on Fuzzy Systems and Knowledge Discovery*, (pp. 620-624). IEEE.

Chen, Y., Dass, S. C., & Jain, A. K. (2005). Fingerprint quality indices for predicting authentication performance. In *Proceedings of Fifth International Conference on Audio and Video-Based Biometric Person Authentication (AVBPA)*, (pp. 373-381). Rye Brook, NY: AVBPA.

Chen, C., Liang, J., & Zhao, H. (2009). Frame difference energy image for gait recognition with incomplete silhouettes. *Pattern Recognition Letters*, *30*, 977–984. doi:10.1016/j.patrec.2009.04.012.

Chen, H., & Jain, A. K. (2005). Dental biometrics: Alignment and matching of dental radiographs. *IEEE Transactions on Pattern Analysis and Machine Intelligence*, *27*(8), 1319–1326. doi:10.1109/TPAMI.2005.157 PMID:16119269.

Chen, L., & Aihara, K. (1995). Chaotic simulated annealing by a neural network model with transient chaos. *Neural Networks*, *8*(6), 915–930. doi:10.1016/0893-6080(95)00033-V.

Chen, L., & Aihara, K. (1997). Chaos and asymptotical stability in discrete time neural networks. *Physica D. Nonlinear Phenomena*, *104*, 286–325. doi:10.1016/S0167-2789(96)00302-8.

Choi, M. Y., & Huberman, B. A. (1983). Dynamic behavior of nonlinear networks. *Physical Review A.*, *28*, 1204–1206. doi:10.1103/PhysRevA.28.1204.

Cole, J. (2008). *Second life salon*. Retrieved from http://www.salon.com/opinion/feature/2008/02/25/avatars/

Condorcet, M.-J. (1785). *Essai sur l'application de l'analyse a la probabilite des decisions rendues a la pluralite des voix*. Paris, France: Academic Press.

Copeland, A. H. (1951). *A reasonable social welfare function*. Ann Arbor, MI: University of Michigan.

Corney, M., Vel, O. D., Anderson, A., & Mohay, G. (2002). Gender-preferential text mining of e-mail discourse. In *Proceedings of 18th Annual Computer Security Applications Conference*, (pp. 282-289). Brisbane, Australia: IEEE.

Craig, N. L., Cohen-Fix, O., Green, R., Greider, C. W., Storz, G., & Wolberger, C. (2011). *Molecular biology principles of genome function*. Oxford, UK: Oxford University Press.

Craw, I., Tock, D., & Bennett, A. (1992). Finding face features. In *Proceedings of Second European Conference on Computer Vision*, (pp. 92-96). Santa Margherita Ligure, Italy: IEEE.

Cummins, H., & Kennedy, R. (1940). Purkinji's observations (1823) on fingerprints and other skin features. *The American Journal of Police Science, 31*(3).

Cuntoor, N., Kale, A., & Chellappa, R. (2003). Combining multiple evidences for gait recognition. In *Proceedings of the 2003 International Conference on Multimedia and Expo*, (pp. 113-116). IEEE Computer Society.

Cutting, J. K., & Kozlowski, L. K. (1977). Recognizing friends by their walk: Gait perception without familiarity cues. *Bulletin of the Psychonomic Society, 9*(5), 353–356.

Dai, Y., & Nakano, Y. (1996). Face-texture model based on SGLD and its application in face detection in a color scene. *Pattern Recognition, 29*(6), 1007–1017. doi:10.1016/0031-3203(95)00139-5.

Daleno1, D., Dellisanti, M., & Giannini, M. (2008). Retinal fundus biometric analysis for personal identifications. In *Proceedings of ICIC*, (pp. 1229-1237). Springer.

Daugman, J. (2000). *Combining multiple biometrics*. Retrieved from http://www.cl.cam.ac.uk/users/jgd1000/combine/combine.html

Daugman, J. G., & Williams, G. O. (1996). A proposed standard for biometric decidability. In Proceedings of cardTechSecureTech, (pp. 223-224). Atlanta, GA: cardTechSecureTech.

Daugman, J. (1993). High confidence visual recognition of persons by a test of statistical independence. *IEEE Transactions on Pattern Analysis and Machine Intelligence, 15*, 1148–1161. doi:10.1109/34.244676.

Daugman, J. (2004). How iris recognition works. *IEEE Transactions on Circuits and Systems for Video Technology, 14*(1), 21–30. doi:10.1109/TCSVT.2003.818350.

Delhi, I. I. T. (2012). *Ear database*. Retrieved from http://www4.comp.polyu.edu.hk/~csajaykr/IITD/Database_Ear.htm

Dempster, A. P. (1976). Upper and lower probabilities induced by a multivalued mapping. *Annals of Mathematical Statistics, 38*(2), 325–339. doi:10.1214/aoms/1177698950.

Deng, Y., Su, X., Wang, D., & Li, Q. (2010). Target recognition based on fuzzy Dempster data fusion method. *Defence Science Journal, 60*(5), 525–530.

Dobeš, M., Martinek, J., Skoupil, D., Dobešová, Z., & Pospíšil, J. (2006). Human eye localization using the modified Hough transform. *Optik (Stuttgart), 117*(10), 468–473. doi:10.1016/j.ijleo.2005.11.008.

Dunstone, T., & Yager, N. (2006). *Biometric system and data analysis: Design, evaluation, and data mining*. New York, NY: Springer.

Dutkowski, J., & Gambin, A. (2007). On consensus biomarker selection. *BioMed Central Bioinformatics, 24*(8), S5. doi:10.1186/1471-2105-8-S5-S5 PMID:17570864.

Dwork, C., Kumar, R., Naor, M., & Sivakumar, D. (2001). Rank aggregation methods for the web. In *Proceedings of Tenth International Conference on the World Wide Web (WWW)*, (pp. 613-622). Hong Kong, China: IEEE.

Egan, J. (1975). *Signal detection theory and ROC analysis*. New York, NY: Academic Press.

Eisenberg, J., Freeman, W. J., & Burke, B. (1989). Hardware architecture of a neural network model simulating pattern recognition by the olfactory bulb. *Neural Networks, 2*, 315–325. doi:10.1016/0893-6080(89)90040-3.

Ester, M., Kriegel, H. P., Sander, J., & Xu, X. (1996). A density-based algorithm for discovering clusters in large spatial databases with noise. In *Proceedings of 2nd International Conference on Knowledge Discovery*, (pp. 226-231). IEEE.

Fagin, R. (1999). Combining fuzzy information from multiple systems. *Journal of Computer and System Sciences, 58*(1), 83–99. doi:10.1006/jcss.1998.1600.

Farah, M., & Vanderpooten, D. (2008). An outranking approach for information retrieval. *Information Retrieval, 11*(4), 315–334. doi:10.1007/s10791-008-9046-z.

Fassinut-Mombot, B., & Choquel, J. B. (2004). A new probabilistic and entropy fusion approach for management of information sources. *Information Fusion*, *5*, 35–47. doi:10.1016/j.inffus.2003.06.001.

Feng, G., Dong, K., Hu, D., & Zhang, D. (2004). When faces are combined with palmprint: A novel biometric fusion strategy. In *Proceedings of First International Conference on Biometric Authentication*, (pp. 701-707). Hong Kong, China: IEEE.

Fishburn, P. C. (1990). A note on "A Note on Nanson's Rule". *Public Choice*, *64*(1), 101–102. doi:10.1007/BF00125920.

Fisher, R. (2012). Avatars can't hide your lying eyes. *New Scientist*. Retrieved from www.newscientist.com/article/mg20627555.600-avatars-cant-hide-your-lying-eyes.html

Fong, T. W., Nourbakhsh, I., & Dautenhahn, K. (2003). A survey of socially interactive robots. *Robotics and Autonomous Systems*, *42*, 143–166. doi:10.1016/S0921-8890(02)00372-X.

Frantzeskou, G., Gritzalis, S., & MacDonell, S. (2004). Source code authorship analysis for supporting the cybercrime investigation process. In *Proceedings of 1st International Conference on eBusiness and Telecommunication Networks - Security and Reliability in Information Systems and Networks Track*, (pp. 85-92). Setubal, Portugal: IEEE.

Freeman, L. C. (1979). Centrality in social networks: Conceptual clarification. *Social Networks*, *1*, 215–239. doi:10.1016/0378-8733(78)90021-7.

Freeman, W. J., & Yao, Y. (1990). Model of biological pattern recognition with spatially chaotic dynamics. *Neural Networks*, *3*, 153–170. doi:10.1016/0893-6080(90)90086-Z.

Frischholz, R., & Dieckmann, U. (2000). BioID: A multimodal biometric identification system. *IEEE Computer*, *33*(2), 64–68. doi:10.1109/2.820041.

Fukai, T., & Shiino, M. (1990). Asymmetric neural networks incorporating the dale hypothesis and noise-driven chaos. *Physical Review Letters*, *64*, 1465–1468. doi:10.1103/PhysRevLett.64.1465 PMID:10041402.

Gabor, D. (2012). Theory of communication. *Journal of the Institute of Electrical Engineering*, *93*, 429–457.

Gambin, A., & Pokarowski, P. (2001). A combinatorial aggregation algorithm for stationary distribution of a large Markov chain. *Lecture Notes in Computer Science*, *2138*, 384–387. doi:10.1007/3-540-44669-9_38.

Gao, W., Cao, B., Shan, S., Chen, X., Zhou, D., Zhang, X., & Zhao, D. (2008). The CAS-PEAL large-scale Chinese face database and baseline evaluations. *IEEE Transactions on Systems, Man and Cybernetics. Part A*, *38*(1), 149–161.

Garris, M. D., Watson, C. I., & Wilson, C. L. (2004). *Matching performance for the US-visit IDENT system using flat fingerprints*. Technical Report 7110. Washington, DC: National Institute of Standards and Technology (NIST).

Gavrilova, M. (2004). Computational geometry and biometrics: On the path to convergence. In *Proceedings of the International Workshop on Biometric Technologies 2004*, (pp. 131-138). Calgary, Canada: IEEE.

Gavrilova, M., & Yampolskiy, R. (2012). Applying biometric principles to avatar recognition. In *Proceedings of IEEE RAM*. IEEE.

Gavrilova, L., & Yampolskiy, R. V. (2011). Applying biometric principles to avatar recognition. *Transactions on Computational Science*, *12*, 140–158. doi:10.1007/978-3-642-22336-5_8.

Gavrilova, M., & Ahmadian, K. (2011). Dealing with biometric multi-dimensionality through novel chaotic neural network methodology. *International Journal of Information Technology and Management*, *11*(1-2), 18–34.

Gavrilova, M., & Monwar, M. M. (2008). Fusing multiple matcher's outputs for secure human identification. *International Journal of Biometrics*, *1*(3), 329–348. doi:10.1504/IJBM.2009.024277.

Gavrilova, M., & Monwar, M. M. (2011). Current trends in multimodal system development: Rank level fusion. In Wang, P. (Ed.), *Pattern Recognition, Machine Intelligence and Biometrics (PRMIB)*. Berlin, Germany: Springer. doi:10.1007/978-3-642-22407-2_25.

Gianvecchio, S., Xie, M., Wu, Z., & Wang, H. (2008). Measurement and classification of humans and bots in internet chat. In *Proceedings of 17th Conference on Security Symposium*, (pp. 155-169). San Jose, CA: IEEE.

Golfarelli, M., Maio, D., & Maltoni, D. (1997). On the error-reject tradeoff in biometric verification systems. *IEEE Transactions on Pattern Analysis and Machine Intelligence, 19*(7), 786–796. doi:10.1109/34.598237.

Gray, A., Sallis, P., & MacDonell, S. (1997). Software forensics: Extending authorship analysis techniques to computer programs. In *Proceedings of 3rd Biannual Conference of the International Association of Forensic Linguists*. IEEE.

Griaule Biometrics. (2012). *Website.* Retrieved online from http://www.griaulebiometrics.com/en-us/book/understanding-biometrics/introduction/types/behavioral

Grinstead, C. M., & Snell, J. L. (1997). *Introduction to probability* (2nd ed.). Providence, RI: American Mathematical Society.

Halteren, H. V. (2004). Linguistic profiling for author recognition and verification. In *Proceedings of ACL*. ACL.

Hamming, R. W. (1950). Error detecting and error correcting codes. *The Bell System Technical Journal, 29*(2), 147–160.

Han, J., & Bhanu, B. (2006). Individual recognition using gait energy image. *IEEE Transactions on Pattern Analysis and Machine Intelligence, 28*(2), 316–322. doi:10.1109/TPAMI.2006.38 PMID:16468626.

Han, J., & Kamber, M. (2001). *Data mining concepts and techniques.* San Francisco, CA: Kaufmann.

Harb, A. M., & Al-Smadi, I. (2006). Chaos control using fuzzy controllers (Mamdani model), integration of fuzzy logic and chaos theory. *Studies in Fuzziness and Soft Computing, 187,* 127–155. doi:10.1007/3-540-32502-6_6.

He, M. et al. (2010). Performance evaluation of score level fusion in multimodal biometric systems. *Pattern Recognition, 43*(5), 1789–1800. doi:10.1016/j.patcog.2009.11.018.

Holz, T., Dragone, M., & O'Hare, G. P. (2009). Where robots and virtual agents meet: A survey of social interaction research across Milgram's reality-virtuality continuum. *International Journal of Social Robotics, 1*(1). doi:10.1007/s12369-008-0002-2.

Hong, L., & Jain, A. K. (1998). Integrating faces and fingerprints for personal identification. *IEEE Transactions on Pattern Analysis and Machine Intelligence, 20*(12), 1295–1307. doi:10.1109/34.735803.

Hopfield, J. J. (1990). The effectiveness of analogue neural network hardware. *Network (Bristol, England), 1*(1), 27–40. doi:10.1088/0954-898X/1/1/003.

Ho, T. K., Hull, J. J., & Srihari, S. N. (1994). Decision combination in multiple classifier systems. *IEEE Transactions on Pattern Analysis and Machine Intelligence, 16*(1), 66–75. doi:10.1109/34.273716.

Hough, P. V. C. (1962). *Method and means for recognizing complex patterns.* US Patent 3069654. Washington, DC: US Patent Office.

Hsu, R.-L. (2002). *Face detection and modeling for recognition.* (PhD Thesis). Michigan State University. East Lancing, MI.

Hunt, K. J. (1996). Extending the functional equivalence of radial basis function networks and fuzzy inference system. *IEEE Transactions on Neural Networks, 13,* 776–778. doi:10.1109/72.501735.

Iannarelli, A. (1989). *Ear identification.* Fremont, CA: Paramont Publishing Company.

Iris Recognition. (2003). *Iris technology division.* Cranbury, NJ: LG Electronics USA.

Ito, J. (2009). Fashion robot to hit Japan catwalk. *PHYSorg.* Retrieved from www.physorg.com/pdf156406932.pdf.

Jain, A. Flynn, P., & Ross, A. (2007). Handbook of biometrics. New York, NY: Springer.

Jain, A. K., Ross, A., & Pankanti, S. (1999). A prototype hand geometry-based verification system. In *Proceedings of International Conference on Audio- and Video-based Biometric Person Authentication*, (pp. 166-171). IEEE.

Jain, A. K. (2005). Biometric recognition: How do I know who you are? *Lecture Notes in Computer Science, 3617,* 19–26. doi:10.1007/11553595_3.

Jain, A. K., Bolle, R., & Pankanti, S. (Eds.). (1999). *Biometrics: Personal identification in networked society.* Dordrecht, The Netherlands: Kluwer Academic Publishers.

Jain, A. K., Flynn, P., & Ross, A. (2007). *Handbook of biometrics.* New York, NY: Springer.

Jain, A. K., Hong, L., & Bolle, R. (1997). On-line fingerprint verification. *IEEE Transactions on Pattern Analysis and Machine Intelligence, 19*(4), 302–314. doi:10.1109/34.587996.

Jain, A. K., & Li, S. Z. (2004). *Handbook on face recognition.* New York, NY: Springer-Verlag.

Jain, A. K., Nandakumar, K., & Ross, A. (2005). Score normalization in multimodal biometric systems. *Pattern Recognition, 38,* 2270–2285. doi:10.1016/j.patcog.2005.01.012.

Jain, A. K., Ross, A., & Prabhakar, A. (2004). An introduction to biometric recognition. *IEEE Transactions on Circuits and Systems for Video Technology, 14*(1), 4–20. doi:10.1109/TCSVT.2003.818349.

Jain, A., Hong, L., & Bolle, R. (1997). On-line fingerprint verification. *IEEE Transactions on Pattern Analysis and Machine Intelligence, 4,* 302–313. doi:10.1109/34.587996.

Jain, A., & Kumar, A. (2012). Biometric recognition: An overview. *The International Library of Ethics. Law and Technology, 11,* 49–79.

Jiang, X., & Yau, W.-Y. (2000). Fingerprint minutiae matching based on the local and global structures. In *Proceedings of the 15th Internet Conference on Pattern Recognition (ICPR, 2000),* (vol. 2, pp. 1042–1045). ICPR.

Jing, X. Y., Yao, Y. F., Yang, J. Y., Li, M., & Zhang, D. (2007). Face and palmprint pixel level fusion and kernel DCV-RBF classifier for small sample biometric recognition. *Pattern Recognition, 40,* 3209–3224. doi:10.1016/j.patcog.2007.01.034.

Joachim, D., Jorg, K., Edda, L., & Paass, G. (2003). Authorship attribution with support vector machines. In *Proceedings of Applied Intelligence* (pp. 109–123). IEEE.

Johansson, G. (1973). Visual perception of biological motion and a model for its analysis. *Perception & Psychophysics, 14*(2), 201–211. doi:10.3758/BF03212378.

Johnson, R. G. (1991). *Can iris patterns be used to identify people?* Los Alamos, CA: Chemical and Laser Sciences Division Los Alamos National Laboratory.

Juola, P., & Sofko, J. (2004). Proving and improving authorship attribution. In *Proceedings of CaSTA.* CaSTA.

Kanda, T., Ishiguro, H., Ono, T., Imai, M., & Mase, K. (2002). Multi-robot cooperation for human-robot communication. In *Proceedings of 11th IEEE International Workshop on Robot and Human Interactive Communication,* (pp. 271- 276). IEEE.

Kemeny, J. G., Snell, J. L., & Thompson, G. L. (1974). *Introduction to finite mathematics* (3rd ed.). Englewood Cliffs, NJ: Prentice-Hall.

Khurshid, J., & Bing-Rong, H. (2004). Military robots - A glimpse from today and tomorrow. In *Proceedings of 8th Control, Automation, Robotics and Vision Conference,* (pp. 771-777). IEEE.

Kim, K. I., Jung, K., & Kim, H. J. (2002). Face recognition using kernel principal component analysis. *IEEE Signal Processing Letters, 9*(2), 40–42. doi:10.1109/97.991133.

Kim, S.-W., Kim, K., Lee, J.-H., & Cho, D. (2001). Application of fuzzy logic to vehicle classification algorithm in Loop/Piezo-sensor fusion systems. *Asian Journal of Control, 3*(1), 64–68. doi:10.1111/j.1934-6093.2001.tb00044.x.

Kittler, J., Hatef, M., Duin, R. P., & Matas, J. G. (1998). On combining classifiers. *IEEE Transactions on Pattern Analysis and Machine Intelligence, 20*(3), 226–239. doi:10.1109/34.667881.

Kjell, B. (1994). Authorship attribution of text samples using neural networks and Bayesian classifiers. In *Proceedings of, IEEE International Conference on Systems, Man, and Cybernetics. 'Humans, Information and Technology'*, (pp. 660-1664). San Antonio, TX: IEEE.

Klimpak, B., Grgic, M., & Delac, K. (2006). Acquisition of a face database for video surveillance research. In *Proceedings of 48th International Symposium focused on Multimedia Signal Processing and Communications*, (pp. 111-114). IEEE.

Klingspor, V., Demiris, J., & Kaiser, M. (1997). Human-robot-communication and machine learning. *Applied Artificial Intelligence*, *11*, 719–746.

Kludas, J., Bruno, E., & Marchand-Maillet, S. (2008). Information fusion in multimedia information retrieval. *Lecture Notes in Computer Science*, *4918*, 147–159. doi:10.1007/978-3-540-79860-6_12.

Kobayashi, H., & Hara, F. (1993). Study on face robot for active human interface-mechanisms of facerobot and expression of 6 basic facial expressions. In *Proceedings of 2nd IEEE International Workshop on Robot and Human Communication*, (pp. 276-281). Tokyo, Japan: IEEE.

Kokar, M. M., Weyman, J., & Tomasik, J. A. (2004). Formalizing classes of information fusion systems. *Information Fusion*, *5*, 189–202. doi:10.1016/j.inffus.2003.11.001.

Kolmogorov, A. N. (1957). On the representation of continuous functions of several variables by superposition of continuous functions of one variable and addition. *Doklady Akademii: Nauk USSR*, *114*, 679–681.

Koppel, M., & Schler, J. (2004). Authorship verification as a one-class classification problem. In *Proceedings of 21st International Conference on Machine Learning*, (pp. 489-495). Banff, Canada: IEEE.

Koppel, M., Schler, J., & Mughaz, D. (2004). Text categorization for authorship verification. In *Proceedings of Eighth International Symposium on Artificial Intelligence and Mathematics*. Fort Lauderdale, FL: IEEE.

Koschutzki, D., & Lehmann, K. A. Peeters, L., Richter, S., Tenfelde-Podehl, D., & Zlotowski, O. (2005). Centrality indices. In Proceedings of Network Analysis, (pp. 16–61). Berlin, Germany: Springer.

Kovacs-Vajna, Z., & Miklos, A. (2000). Fingerprint verification system based on triangular matching and dynamic time warping. *IEEE Transactions on Pattern Analysis and Machine Intelligence*, *22*(11), 1266–1276. doi:10.1109/34.888711.

Kumar, A., & Shekhar, S. (2010). Palmprint recognition using rank level fusion. In *Proceedings of IEEE International Conference on Image Processing*, (pp. 3121-3124). Hong Kong, China: IEEE.

Kuncheva, L. I. (2004). *Combining pattern classifiers: Methods and algorithms*. New York, NY: Wiley. doi:10.1002/0471660264.

Kung, S. Y., Mak, M. W., & Lin, S. H. (2005). *Biometric authentication: A machine learning approach*. Upper Saddle River, NJ: Prentice Hall.

Lam, T. H. W., & Lee, R. S. T. (2006). A new representation for human gait recognition: Motion silhouettes image (MSI). In *Proceedings of the International Conference on Biometrics*, (pp. 612–618). IEEE.

Lam, L., & Suen, C. Y. (1995). Optimal combination of pattern classifiers. *Pattern Recognition Letters*, *16*, 945–954. doi:10.1016/0167-8655(95)00050-Q.

Lam, L., & Suen, C. Y. (1997). Application of majority voting to pattern recognition: An analysis of its behavior and performance. *IEEE Transactions on Systems, Man, and Cybernetics. Part A, Systems and Humans*, *27*(5), 553–568. doi:10.1109/3468.618255.

Lanitis, A., Taylor, C. J., & Cootes, T. F. (1995). An automatic face identification system using flexible appearance model. *Image and Vision Computing*, *13*(5), 393–401. doi:10.1016/0262-8856(95)99726-H.

Lee, M. W., & Ranganath, S. (2003). Pose-invariant face recognition using a 3D deformable model. *Pattern Recognition*, *36*(8), 1835–1846. doi:10.1016/S0031-3203(03)00008-6.

Lim, H.-O., & Takanishi, A. (2000). Waseda biped humanoid robots realizing human-like motion. In *Proceedings of 6th International Workshop on Advanced Motion Control*, (pp. 525-530). Nagoya, Japan: IEEE.

Linas, J., Bowman, C., Rogova, G., Steinberg, A., Waltz, E., & White, F. (2004). Revisiting the JDL data fusion model II. In *Proceedings of 7th International Conference on Information Fusion*. Stockholm, Sweden: IEEE.

Lippmann, R. P. (1987). An introduction to computing with neural nets. *IEEE Magazine on Accoustics, Signal, and Speech Processing, 4*, 4–22.

Li, S., & Jain, A. K. (2005). *Handbook of face recognition - Face databases*. New York, NY: Springer.

Liu, J., & Zheng, N. (2007). Gait history image: A novel temporal template for gait recognition. In *Proceedings of the 2007 IEEE International Conf. on Multimedia and Expo*, (pp. 663-666). Beijing, China: IEEE.

Liu, X., & Chen, T. (2003). Geometry-assisted statistical modeling for face mosaicing. In *Proceedings of IEEE International Conference on image Processing*, (vol. 2, pp. 883-886). Barcelona, Spain: IEEE.

Liu, Z., & Sarkar, S. (2004). Simplest representation yet for gait recognition: averaged silhouette. In *Proceedings of the 17th International Conference on Pattern Recognition*, (pp. 211-214). IEEE.

Lu, J., Plataniotis, K. N., & Venetsanopoulos, A. N. (2003). Face recognition using LDA-based algorithms. *IEEE Transactions on Neural Networks, 14*(1), 195–200. doi:10.1109/TNN.2002.806647 PMID:18238001.

Luo, Y., & Gavrilova, M. (2006). 3D facial model synthesis using voronoi approach. In *Proceedings of IEEE ISVD*, (pp. 132-137). Banff, Canada: IEEE.

Lyons, M., Plante, A., Jehan, S., Inoue, S., & Akamatsu, S. (1998). Avatar creation using automatic face recognition. In *Proceedings of ACM Multimedia 98*, (pp. 427-434). Bristol, UK: ACM.

Lyu, M. R., King, I., Wong, T. T., Yau, E., & Chan, P. W. (2005). ARCADE: Augmented reality computing arena for digital entertainment. In *Proceedings of IEEE Aerospace Conference*. Big Sky, MT: IEEE.

Ma, Q., Wang, S., Nie, D., & Qiu, J. (2007). Recognizing humans based on gait moment image. In *Proceedings of the Eighth ACIS International Conference on Software Engineering, Artificial Intelligence, Networking, and Parallel/Distributed Computing*, (pp. 606-610). IEEE Computer Society.

Maltoni, D., Maio, D., Jain, A. K., & Prabhakar, S. (2009). *Handbook of fingerprint recognition* (2nd ed.). New York, NY: Springer-Verlag. doi:10.1007/978-1-84882-254-2.

Markov, A. A. (1906). Extension of the limit theorems of probability theory to a sum of variables connected in a chain. In R. Howard (Ed.), Dynamic Probabilistic Systems, Volume 1: Markov Chains. Hoboken, NJ: John Wiley and Sons.

Martín-Félez, R., Mollineda, R. A., & Sánchez, J. S. (2011). Human recognition based on gait poses. *Lecture Notes in Computer Science, 6669*, 347–354. doi:10.1007/978-3-642-21257-4_43.

Matsumoto, T., Matsumoto, H., Yamada, K., & Hoshino, S. (2002). Impact of artificial 'gummy' fingers on fingerprint systems.[). SPIE.]. *Proceedings of SPIE Optical Security and Counterfeit Deterrence Techniques IV, 4677*, 275–289. doi:10.1117/12.462719.

McCulloch, J. L., & Pitts, W. (1943). A logical calculus of ideas immanent in nervous activity. *The Bulletin of Mathematical Biophysics, 5*, 115–133. doi:10.1007/BF02478259.

McKenna, S., Gong, S., & Raja, Y. (1998). Modelling facial colour and identity with Gaussian mixtures. *Pattern Recognition, 31*, 1883–1892. doi:10.1016/S0031-3203(98)00066-1.

Mishra, P., & Erza, S. (2010). Human gait recognition using Bezier curves. *International Journal on Computer Science and Engineering, 3*(2), 969–975.

Misra, D., & Gaj, K. (2006). Face recognition CAPTCHAs. In *Proceedings of International Conference on Telecommunications, Internet and Web Applications and Services*. IEEE.

Mitra, S., & Pal, S. K. (2005). Fuzzy sets in pattern recognition and machine intelligence. *Fuzzy Sets and Systems, 156*, 381–386. doi:10.1016/j.fss.2005.05.035.

Mohamed, A., Gavrilova, A., & Yampolskii, R. (2012). Artificial face recognition using wavelet adaptive LBP with directional statistical features. In *Proceedings of CyberWorlds 2012*. IEEE. doi:10.1109/CW.2012.11.

Monwar, M. M., & Gavrilova, M. (2008). FES: A system of combining face, ear and signature biometrics using rank level fusion. In *Proceedings of 5th International Conference on Information Technology: New Generations*, (pp. 922-927). Las Vegas, NV: IEEE.

Monwar, M. M., & Gavrilova, M. (2010). Secured access control through Markov chain based rank level fusion method. In *Proceedings of the 5th International Conference on Computer Vision Theory and Applications (VISAPP)*, (pp. 458-463). Angers, France: VISAPP.

Monwar, M. M., & Gavrilova, M. (2011). Markov chain model for multimodal biometric rank fusion. In Proceedings of Signal, Image and Video Processing, (pp. 1-13). Springer-Verlag.

Monwar, M., Gavrilova, M., & Wang, Y. (2011). A novel fuzzy multimodal information fusion technology for human biometric traits identification. In *Proceedings of ICCI*CC*. Banff, Canada: IEEE.

Monwar, M. M., & Gavrilova, M. (2009). A multimodal biometric system using rank level fusion approach. *IEEE Transactions on Systems, Man, and Cybernetics. Part B, Cybernetics, 39*(4), 867–878. doi:10.1109/TSMCB.2008.2009071 PMID:19336340.

Monwar, M. M., & Gavrilova, M. (2011). Markov chain model for multimodal biometric rank fusion. In *Proceedings of Signal, Image and Video Processing*. Springer. doi:10.1007/s11760-011-0226-8.

Moon, Y. S., Yeung, H. W., Chan, K. C., & Chan, S. O. (2004). Template synthesis and image mosaicing for fingerprint registration: An experimental study. In *Proceedings of IEEE International Conference on Acoustics, Speech, and Signal Processing*, (vol. 5, pp. 409-412). Montreal, Canada: IEEE.

Moon, H., & Phillips, P. J. (2001). Computational and performance aspects of PCA-based face recognition algorithms. *Perception, 30*(5), 303–321. doi:10.1068/p2896 PMID:11374202.

Moreno, J. L. (1934). *Who shall survive?* Washington, DC: Nervous and Mental Disease Publishing Company.

Moustakas, K. T., & Stavropoulos, D. G. (2010). Gait recognition using geometric features and soft biometrics. *IEEE Signal Processing Letters, 17*(4), 367–370. doi:10.1109/LSP.2010.2040927.

Nandakumar, K., Chen, Y., Dass, S. C., & Jain, A. K. (2009). Likelihood ratio-based biometric score fusion. *IEEE Transactions on Pattern Analysis and Machine Intelligence, 30*(2), 342–347. doi:10.1109/TPAMI.2007.70796 PMID:18084063.

Naor, M. (1996). *Verification of a human in the loop or identification via the turing test*. Rehovot, Israel: Weizmann Institute of Science.

Nood, D. D., & Attema, J. (2009). *The second life of virtual reality*. Retrieved from http://www.epn.net/interrealiteit/EPN-REPORT-The_Second_Life_of_VR.pdf

Nozawa, H. (1992). A neural-network model as a globally coupled map and applications based on chaos. *Chaos (Woodbury, N.Y.), 2*(3), 377–386. doi:10.1063/1.165880 PMID:12779987.

Oh, J.-H., Hanson, D., Kim, W.-S., Han, I. Y., Han, Y., & Park, I.-W. (Eds.). (2006). *Proceedings of international conference on intelligent robots and systems*. Daejeon, South Korea: IEEE.

Osuna, E., Freund, R., & Girosi, F. (1997). Training support vector machines: An application to face detection. In *Proceedings of IEEE Conference on Computer Vision and Pattern Recognition*, (pp. 130-136). IEEE.

Oursler, J. N., Price, M., & Yampolskiy, R. V. (2009). Parameterized generation of avatar face dataset. In *Proceedings of 14th International Conference on Computer Games: AI, Animation, Mobile, Interactive Multimedia, Educational & Serious Games*. Louisville, KY: IEEE.

Oviatt, S. (2003). Advances in robust multimodal interface design. *IEEE Computer Graphics and Applications, 23*(5), 62–88. doi:10.1109/MCG.2003.1231179.

Patel, P., & Hexmoor, H. (2009). Designing BOTs with BDI agents. In *Proceedings of International Symposium on Collaborative Technologies and Systems (CTS)*. (pp. 180-186). Carbondale, PA: IEEE.

Paul, P. P., & Gavrilova, M. (2012). Multimodal cancellable biometric. In *Proceedings of the 10th IEEE International Conference on Cognitive Informatics & Cognitive Computing (ICCI*CC)*, (pp. 43-50). IEEE.

Paul, P. P., Monwar, M., Gavrilova, M., & Wang, P. (2010). Rotation invariant multi-view face detection using skin color regressive model and support vector regression. *International Journal of Pattern Recognition and Artificial Intelligence, 24*(8), 1261–1280. doi:10.1142/S0218001410008391.

Pedrycz, W., & Gomide, F. A. C. (1998). *An introduction to fuzzy sets: Analysis and design complex adaptive systems*. Cambridge, MA: MIT Press.

Pennock, D. M., & Horvitz, E. (2000). Social choice theory and recommender systems: Analysis of the axiomatic foundations of collaborative filtering. In *Proceedings of Seventeenth National Conference on Artificial Intelligence and Twelfth Conference on Innovative Applications of Artificial Intelligence*, (pp. 729-734). Austin, TX: IEEE.

Perpinan, C. (1995). *Compression neural networks for feature extraction: Application to human recognition from ear images*. (M.Sc. Thesis). Technical University of Madrid. Madrid, Spain.

Phillips, P. J., Flynn, P. J., Scruggs, T., Bowyer, K. W., Chang, J., & Hoffman, K. Worek, W. (2005). Overview of the face recognition grand challenge. In *Proceedings of IEEE Computer Society Conference on Computer Vision and Pattern Recognition*, (pp. 947-954). San Diego, CA: IEEE.

Phillips, P. J. (1998). Support vector machines applied to face recognition. *Advances in Neural Information Processing Systems, 11*, 113–123.

Phillips, P. J., Moon, H., & Rauss, P. (1998). The FERET database and evaluation procedure for face recognition algorithms. *Image and Vision Computing, 16*(5), 295–306. doi:10.1016/S0262-8856(97)00070-X.

Pihur, V., Datta, S., & Datta, S. (2008). Finding cancer genes through meta-analysis of microarray experiments: Rank aggregation via the cross entropy algorithm. *Genomics, 92*, 400–403. doi:10.1016/j.ygeno.2008.05.003 PMID:18565726.

Poh, N., & Bengio, S. (2005). How do correlation and variance of base-experts affect fusion in biometric authentication tasks? *IEEE Transactions on Acoustics, Speech, and Signal Processing, 53*, 4384–4396. doi:10.1109/TSP.2005.857006.

Preparata, F., & Shamos, M. (1985). *Computational geometry: An introduction*. Berlin, Germany: Springer. doi:10.1007/978-1-4612-1098-6.

Putte, T., & Keuning, J. (2000). Biometrical fingerprint recognition: Don't get your fingers burned. In *Proceedings of IFIP TC8/WG8.8 Fourth Working Conference on Smart Card Research and Advanced Applications*, (pp. 289-303). Bristol, UK: IFIP.

Raghavendra, R., Rao, A., & Kumar, G. H. (2010). Multi-sensor biometric evidence fusion of face and palmprint for person authentication using particle swarm optimization (PSO). *International Journal of Biometrics, 2*(1), 19–33. doi:10.1504/IJBM.2010.030414.

Rajagopalan, A., Kumar, K., Karlekar, J., Manivasakan, R., Patil, M., & Desai, U. Chaudhuri, S. (1998). Finding faces in photographs. In *Proceedings of, 6th IEEE Intern. Conference on Computer Vision*, (pp. 640-645). IEEE.

Ratha, N., Senior, A., & Bolle, R. (2001). Tutorial on automated biometrics. In *Proceedings of International Conference on Advances in Pattern Recognition*. Rio de Janeiro, Brazil: IEEE.

Ratha, N. K., Karu, K., Chen, S., & Jain, A. (1996). A real-time matching system for large fingerprint databases. *IEEE Transactions on Pattern Analysis and Machine Intelligence, 18*(8), 799–813. doi:10.1109/34.531800.

Rattani, A., Kisku, D. R., Bicego, M., & Tistarelli, M. (2010). Feature level fusion of face and fingerprint biometrics. In *Proceedings of 1st IEEE International Conference on Biometrics: Theory, Applications and Systems*. Washington, DC: IEEE.

Raupp, S., & Thalmann, D. (2001). Hierarchical model for real time simulation of virtual human crowds. *IEEE Transactions on Visualization and Computer Graphics*, 7(2), 152–164. doi:10.1109/2945.928167.

Renesse, R. L. V. (2002). Implications of applying biometrics to travel documents.[). Springer.]. *Proceedings of the Society for Photo-Instrumentation Engineers*, 4677, 290–298. doi:10.1117/12.462720.

Ross, A. (2007). An introduction to multibiometrics. In *Proceedings of 15th European Signal Processing Conference*. Poznan, Poland: IEEE.

Ross, A., & Govindarajan, R. (2005). Feature level fusion using hand and face biometrics. In *Proceedings of SPIE Conference on Biometric Technology for Human Identification II*, (pp. 196-204). Orlando, FL: SPIE.

Ross, A., & Jain, A. K. (2004). Multimodal biometrics: An overview. In *Proceedings of 12th European Signal Processing Conference*, (pp. 1221-1224). Vienna, Austria: IEEE.

Ross, A. A., Nandakumar, K., & Jain, A. K. (2006). *Handbook of multibiometric*. Berlin, Germany: Springer.

Ross, A., & Jain, A. K. (2003). Information fusion in biometrics. *Pattern Recognition Letters*, 24, 2115–2125. doi:10.1016/S0167-8655(03)00079-5.

Ross, A., Nandakumar, K., & Jain, A. K. (2006). *Handbook of multibiometrics*. New York, NY: Springer-Verlag.

Rowley, H. A., Baluja, S., & Kanade, T. (1998). Neural network-based face detection. *IEEE Transactions on Pattern Analysis and Machine Intelligence*, 20(1), 23–38. doi:10.1109/34.655647.

Sanderson, C., & Paliwal, K. K. (2001). Information fusion for robust speaker verification. In *Proceedings of Seventh European Conference on Speech Communication and Technology*, (pp. 755-758). Alborg, Denmark: IEEE.

Sandler, & Yu, M. (1990). Model of neural networks with selective memorization and chaotic behavior. *Physics Letters A*, 144, 462-466.

Sarkar, S., Phillips, P. J., Liu, Z., Vega, I. R., Grother, P., & Bowyer, K. W. (2005). The HumanID gait challenge problem: Data sets, performance, and analysis. *IEEE Transactions on Pattern Analysis and Machine Intelligence*, 27(2), 162–177. doi:10.1109/TPAMI.2005.39 PMID:15688555.

Schneiderman, H., & Kanade, T. (1998). Probabilistic modeling of local appearance and spatial relationships for object recognition. In *Proceedings of IEEE Conference on Computer Vision and Pattern Recognition*, (pp. 45-51). IEEE.

Sculley, D. (2006). Rank aggregation for similar items. Report. New York, NY: Data Mining and Research group of Yahoo.

Shafer, G. (1976). *A mathematical theory of evidence*. Princeton, NJ: Princeton University Press.

Sharma, S., Tiwari, R., Shukla, A., & Singh, V. (2011). Frontal view gait based recognition using PCA. In *Proceedings of the International Conference on Advances in Computing and Artificial Intelligence*, (pp. 124-127). ACM.

Shrotri, A., Rethrekar, S. C., Patil, M. H., Bhattacharyya, D., & Kim, T.-H. (2009). Infrared imaging of hand vein patterns for biometric purposes. *Journal of Security Engineering*, 2, 57–66.

Sim, T., Baker, S., & Bsat, M. (2003). The CMU pose, illumination, and expression database. *IEEE Transactions on Pattern Analysis and Machine Intelligence*, 25(12), 1615–1618. doi:10.1109/TPAMI.2003.1251154.

Singh, R. (2008). *Mitigating the effect of covariates in face recognition*. (PhD Dissertation). University of West Virginia. Morgantown, WV.

Sino Biometrics. (2004). *CASIA: Casia iris image database*. Retrieved from www.sinobiometrics.com

Skarda, C. A., & Freeman, W. J. (1987). How brains make chaos in order to make sense of the world. *The Behavioral and Brain Sciences*, *10*, 161–195. doi:10.1017/S0140525X00047336.

Solaiman, B., Pierce, L. E., & Ulaby, F. T. (1999). Multisensor data fusion using fuzzy concepts: Application to land-cover classification using ERS-1/JERS-1 SAR composites. *IEEE Transactions on Geoscience and Remote Sensing*, *37*, 1316–1326. doi:10.1109/36.763295.

Soledek, J., Shmerko, V., Phillips, P., Kukharevl, G., Rogers, W., & Yanushkevich, S. (1997). Image analysis and pattern recognition in biometric technologies. In *Proceedings of International Conference on the Biometrics: Fraud Prevention, Enhanced Service*, (pp. 270-286). Las Vegas, NV: IEEE.

Soltane, M., Doghmane, N., & Guersi, N. (2010). Face and speech based multi-modal biometric authentication. *International Journal of Advanced Science and Technology*, *21*, 41–56.

Sourina, O., Sourin, A., & Kulish, V. (2009). EEG data driven animation and its application.[MIRAGE.]. *Proceedings of MIRAGE*, *2009*, 380–388.

Spafford, E. H., & Weeber, S. A. (1992). Software forensics: Can we track code to its authors? In *Proceedings of 15th National Computer Security Conference*, (pp. 641-650). IEEE.

Spreecher, D. A. M. (1993). A universal mapping for Kolmogorov's superposition theorem. *Neural Networks*, *6*(8), 1089–1094. doi:10.1016/S0893-6080(09)80020-8.

Stamatatos, E., Fakotakis, N., & Kokkinakis, G. (1999). Automatic authorship attribution. In *Proceedings of Ninth Conference of the European Chapter of the Association of Computational Linguistics*, (pp. 158-164). Bergen, Norway: IEEE.

Stamatatos, E., Fakotakis, N., & Kokkinakis, G. (2001). Computer-based authorship attribution without lexical measures. *Computers and the Humanities*, *35*(2), 193–214. doi:10.1023/A:1002681919510.

Stolfo, S. J., Hershkop, S., Wang, K., Nimeskern, O., & Hu, C.-W. (2003). A behavior-based approach to securing email systems. *Mathematical Methods. Models and Architectures for Computer Networks Security*, *2776*, 57–81. doi:10.1007/978-3-540-45215-7_5.

Suler, J. (2009). *The psychology of cyberspace*. Retrieved from http://psycyber.blogspot.com

Swathi, N. (2011). New palmprint authentication system by using wavelet based method. *Signal & Image Processing: An International Journal*, *2*(1), 191–203. doi:10.5121/sipij.2011.2114.

Tanaka, T., & Kubo, N. (2004). Biometric authentication by hand vein patterns. In *Proceedings of SICE Annual Conference*, (pp. 249-253). Sapporo, Japan: SICE.

Tang, H., Fu, Y., Tu, J., Hasegawa-Johnson, M., & Huang, T. S. (2008). Humanoid audio-visual avatar with emotive text-to-speech synthesis. *IEEE Transactions on Multimedia*, *10*, 969–981. doi:10.1109/TMM.2008.2001355.

Tan, X., Chen, S., Zhou, Z.-H., & Zhang, F. (2006). Face recognition from a single image per person: A survey. *Pattern Recognition*, *39*(9), 1725–1745. doi:10.1016/j.patcog.2006.03.013.

Teijido, D. (2009). Information assurance in a virtual world. In *Proceedings of Australasian Telecommunications Networks and Applications Conference*. Canberra, Australia: IEEE.

Thompson, B. G. (2009). *The state of homeland security*. Retrieved from http://hsc-democrats.house.gov/SiteDocuments/20060814122421-06109.pdf

Truchon, M. (1998). *An extension of the condorcet criterion and kemeny orders. Cahier 9813*. Rennes, France: University of Rennes.

Tumer, K., & Gosh, J. (1999). Linear order statistics combiners for pattern classification. In *Proceedings of Combining Artificial Neural Networks* (pp. 127–162). IEEE.

Turing, A. M. (1950). Computing machinery and intelligence. *Mind*, *59*, 433–460. doi:10.1093/mind/LIX.236.433.

Turk, M., & Pentland, A. (1991). Eigenfaces for recognition. *Journal of Cognitive Neuroscience*, *3*(1), 71–86. doi:10.1162/jocn.1991.3.1.71.

Uludag, U., Ross, A., & Jain, A. K. (2004). Biometric template selection and update: A case study in fingerprints. *Pattern Recognition*, *37*(7), 1533–1542. doi:10.1016/j.patcog.2003.11.012.

University of Essex. (2008). *Face database*. Retrieved from http://cswww.essex.ac.uk/mv/allfaces/index.html

USTB. (2012). *Ear database, China*. Retrieved from http://www.ustb.edu.cn/resb/

Vacca, J. R. (2007). *Biometric technologies and verification systems*. Burlington, MA: Butterworth-Heinemann.

Vel, O. D., Anderson, A., Corney, M., & Mohay, G. (2001). Mining email content for author identification forensics. *SIGMOD Record*, *30*(4), 55–64. doi:10.1145/604264.604272.

Veres, G. V., Gordon, L., Carter, J. N., & Nixon, M. S. (2004). What image information is important in silhouette-based gait recognition? In *Proceedings of the 2004 IEEE Conference on Computer Vision and Pattern Recognition*, (pp. 776-782). IEEE.

Verlinde, P., & Cholet, G. (1999). Comparing decision fusion paradigms using k-NN based classifiers, decision trees and logistic regression in a multi-modal identity verification application. In *Proceedings of Second International Conference on Audio- and Video-Based Biometric Person Authentication (AVBPA)*, (pp. 188–193). AVBPA.

von Neumann, J. (1966). *Theory of self-reproducing automate*. Urbana, IL: University of Illinois Press.

Wang, C., & Gavrilova, M. (2004). A multi-resolution approach to singular point detection in fingerprint images. In *Proceedings of the International Conference of Artificial Intelligence*, (vol. 1, pp. 506-511). IEEE.

Wang, C., & Gavrilova, M. (2005). A novel topology-based matching algorithm for fingerprint recognition in the presence of elastic distortions. In *Proceedings of International Conference on Computational Science and its Applications*, (vol. 1, pp. 748-757). Springer.

Wang, C., & Gavrilova, M. (2006). Delaunay triangulation algorithm for fingerprint matching. In *Proceedings of ISVD*, (pp. 208-216). Banff, Canada: ISVD.

Wang, C., Zhang, J., Pu, J., Yuan, X., & Wang, L. (2010). Chrono-gait image: A novel temporal template for gait recognition. In *Proceedings 11th European Conference on Computer Vision*, (pp. 257-270). IEEE.

Wang, H., Gavrilova, M., Luo, Y., & Rokne, J. (2006). An efficient algorithm for fingerprint matching. In *Proceedings of International Conference on Pattern Recognition*, (pp. 1034-1037). IEEE.

Wang, J., She, M., Nahavandi, S., & Kouzani, A. (2010). A review of vision-based gait recognition methods for human identification. In *Proceedings of the 2010 Digital Image Computing: Techniques and Application*, (pp. 320-327). Piscataway, NJ: IEEE.

Wang, Y. (2009). Fuzzy inferences methodologies for cognitive informatics and computational intelligence. In *Proceedings of 8th IEEE International Conference on Cognitive Informatics*, (pp. 241-248). Hong Kong, China: IEEE.

Wang, Y., Berwick, R. C., Haykin, S., Pedrycz, W., Baciu, G., & Bhavsar, V. C. Zhang, D. (2011). Cognitive informatics in year 10 and beyond. In *Proceedings of the 10th IEEE International Conference on Cognitive Informatics & Cognitive Computing (ICCI*CC)*. IEEE.

Wang, Y.-P., Dang, J.-W., Li, Q., & Li, S. (2007). Multimodal medical image fusion using fuzzy radial basis function neural networks. In *Proceedings International Conference on Wavelet Analysis and Pattern Recognition*, (vol. 2, pp. 778-782). Beijing, China: IEEE.

Wang, C.-H. (2005). *A literature survey on human gait recognition techniques. Directed Studies EE8601*. Toronto, Canada: Ryerson University.

Wang, C., Luo, Y., Gavrilova, M. L., & Rokne, J. (2007). Fingerprint image matching using a hierarchical approach. In Nedjah, N., Abraham, A., & de Macedo Mourelle, L. (Eds.), *Computational Intelligence in Information Assurance and Security* (pp. 175–198). Berlin, Germany: Springer. doi:10.1007/978-3-540-71078-3_7.

Wang, L., & Shi, H. (2006). A gradual noisy chaotic neural network for solving the broadcast scheduling problem in packet radio networks. *IEEE Transactions on Neural Networks*, *17*(4), 989–1001. doi:10.1109/TNN.2006.875976 PMID:16856661.

Wang, L., & Smith, K. (1998). On chaotic simulated annealing. *IEEE Transactions on Neural Networks*, *9*, 716–718. doi:10.1109/72.701185 PMID:18252495.

Wang, Y. (2009). Toward a formal knowledge system theory and its cognitive informatics foundations. *IEEE Transactions of Computational Science*, *5*, 1–19.

Wang, Y. (2009). Toward a formal knowledge system theory and its cognitive informatics foundations. *Transactions of Computational Science*, *5*, 1–19.

Wayman, J. L., Jain, A. K., Maltoni, D., & Maio, D. (2005). An introduction to biometric authentication systems. In Wayman, J. L., Jain, A. K., Maltoni, D., & Maio, D. (Eds.), *Biometric Systems: Technology, Design and Performance Evaluation* (pp. 1–20). London, UK: Springer-Verlag.

Wayman, J., Jain, A., Maltoni, D., & Maio, D. (2006). *Biometric systems: Technology, design and performance evaluation*. Berlin, Germany: Springer-Verlag.

Wecker, L., Samavati, F., & Gavrilova, M. (2005). Iris synthesis: A multi-resolution approach. In *Proceedings of 3rd International Conference on Computer Graphics and Interactive Techniques in Australasia and South East Asia*, (pp. 121-125). IEEE.

Werner, C. (2008). *Biometrics: Trading privacy for security*. Retrieved from http://media.wiley.com/product_data/excerpt/26/07645250/0764525026.pdf

Westerveld, T., & de Vries, A. P. (2004). Multimedia retrieval using multiple examples. In *Proceedings of International Conference on Image and Video Retrieval*. IEEE.

Wilson, C. (2010). *Vein pattern recognition: A privacy-enhancing biometric*. Boca Raton, FL: CRC Press. doi:10.1201/9781439821381.

Woodward, J. D. (1997). Biometrics: Privacy's foe or privacy's friend? *IEEE Proceeding, 85*(9), 1480-1492.

Woodward, J. D. Jr, Horn, C., Gatune, G., & Thomas, A. (2003). *Biometrics: A look at facial recognition*. Arlington, VA: Virginia State Crime Commission.

Wu, Y., Chang, K. C.-C., Chang, E. Y., & Smith, J. R. (2004). Optimal multimodal fusion for multimedia data analysis. In *Proceedings of the 12th Annual ACM International Conference on Multimedia,* (pp. 572-579). ACM Press.

Wu, L., Cohen, P. R., & Oviatt, S. L. (2002). From members to team to committee - A robust approach to gestural and multimodal recognition. *IEEE Transactions on Neural Networks*, *13*.

Wu, S., & McClean, S. (2006). Performance prediction of data fusion for information retrieval. *Information Processing & Management*, *42*, 899–915. doi:10.1016/j.ipm.2005.08.004.

Xiao, Y., & Yan, H. (2002). Facial feature location with delaunay triangulation/voronoi diagram calculation. In Feng, D. D., Jin, J., Eades, P., & Yan, H. (Eds.), *Conferences in Research and Practice in Information Technology* (pp. 103–108). ACS.

Xu, S.-L., & Zhang, Q.-J. (2010). Gait recognition using fuzzy principal component analysis. In *Proceedings of the Second IEEE International Conference on e-Business and Information System Security*. IEEE.

Yamada, T., Aihara, K., & Kotani, M. (1993). Chaotic neural networks and the travelling salesman problem. In *Proceedings of International Joint Conference on Neural Networks,* (pp. 1549–1552). IEEE.

Yampolskiy, R. V. (2007). Behavioral biometrics for verification and recognition of AI programs. In *Proceedings of 20th Annual Computer Science and Engineering Graduate Conference (GradConf)*. Buffalo, NY: GradConf.

Yampolskiy, R. V. (2007). Mimicry attack on strategy-based behavioral biometric. In *Proceedings of 5th International Conference on Information Technology: New Generations*, (pp. 916-921). Las Vegas, NV: IEEE.

Yampolskiy, R. V., & Govindaraju, V. (2007). Behavioral biometrics for recognition and verification of game bots. In *Proceedings of the 8th Annual European Game-On Conference on simulation and AI in Computer Games*. Bologna, Italy: IEEE.

Yampolskiy, R. V., & Govindaraju, V. (2007). Embedded non-interactive continuous bot detection. *ACM Computers in Entertainment*, *5*(4), 1–11. doi:10.1145/1324198.1324205.

Yan, R., & Hauptmann, A. G. (2003). The combination limit in multimedia retrieval. In *Proceedings of the Eleventh ACM International Conference on Multimedia*, (pp. 339-342). ACM Press.

Yang, G., & Huang, T. S. (1994). Human face detection in complex background. *Pattern Recognition*, *27*(1), 53–63. doi:10.1016/0031-3203(94)90017-5.

Yang, M.-H., Kriegman, D. J., & Ahuja, N. (2002). Detecting faces in images: A survey. *IEEE Transactions on Pattern Analysis and Machine Intelligence*, *24*(1).

Yang, X., Zhou, Y., Zhang, T., Shu, G., & Yang, J. (2008). Gait recognition based on dynamic region analysis. *Signal Processing*, *88*(9), 2350–2356. doi:10.1016/j.sigpro.2008.03.006.

Yanushkevich, S., Gavrilova, M., Wang, P., & Srihari, S. (2007). *Image Pattern Recognition: Synthesis and Analysis in Biometrics*. New York, NY: World Scientific Publishers.

Yao, Y., Freeman, W. J., Burke, B., & Yang, Q. (1991). Pattern recognition by a distributed neural network: An industrial application. *Neural Networks*, *4*, 103–121. doi:10.1016/0893-6080(91)90036-5.

Yoo, J.-H., & Nixon, M. (2011). Automated markerless analysis of human gait motion for recognition and classification. *ETRI Journal*, *33*(2), 259–266. doi:10.4218/etrij.11.1510.0068.

Yow, K. C., & Cipolla, R. (1997). Feature-based human face detection. *Image and Vision Computing*, *15*(9), 713–735. doi:10.1016/S0262-8856(97)00003-6.

Yu, P., Xu, D., Zhou, H., & Li, H. (2009). Decision fusion for hand biometric authentication. In *Proceedings of IEEE International Conference on Intelligent Computing and Intelligent Systems*, (vol. 4, pp. 486-490). Shanghai, China: IEEE.

Yuan, L., Gavrilova, M., & Wang, P. (2008). Facial metamorphosis using geometrical methods for biometric applications. *International Journal of Pattern Recognition and Artificial Intelligence*, *22*(3), 555–584. doi:10.1142/S0218001408006399.

Zadeh, L. A. (1965). Fuzzy sets. *Information and Control*, *8*, 338–353. doi:10.1016/S0019-9958(65)90241-X.

Zhang, D. (2004). *Palmprint authentication*. Berlin, Germany: Springer.

Zhao, R., & Grosky, W. I. (2002). Narrowing the semantic gap - Improved text-based web document retrieval using visual features. *IEEE Transactions on Multimedia*, *4*(2), 189–200. doi:10.1109/TMM.2002.1017733.

Zhao, W., Chellappa, R., Phillips, P. J., & Rosenfeld, A. (2003). Face recognition: A literature survey. *ACM Computing Surveys*, *35*(4), 399–458. doi:10.1145/954339.954342.

Zykov, V., Mytilinaios, E., Adams, B., & Lipson, H. (2005). Robotics: Self-reproducing machines. *Nature*, *435*, 163–164. doi:10.1038/435163a PMID:15889080.

Related References

To continue our tradition of advancing information science and technology research, we have compiled a list of recommended IGI Global readings. These references will provide additional information and guidance to further enrich your knowledge and assist you with your own research and future publications.

Abi-Char, P. E., El-Hassan, B., & Mokhtari, M. (2010). Privacy and trust issues in context-aware pervasive computing: State-of-the-Art and future directions. In Yan, Z. (Ed.), *Trust modeling and management in digital environments: From social concept to system development* (pp. 352–377). Hershey, PA: IGI Global. doi:10.4018/978-1-61520-682-7.ch015.

Adebowale, O. (2010). Identity Awareness. In Yuzer, T., & Kurubacak, G. (Eds.), *Transformative learning and online education: Aesthetics, dimensions and concepts* (pp. 316–330). Hershey, PA: IGI Global. doi:10.4018/978-1-61520-985-9. ch020.

Agrawal, L., Kumar, A., Nagori, J., & Varma, S. (2012). Data fusion in wireless sensor networks: Classification, techniques, and models. In Lakhtaria, K. (Ed.), *Technological advancements and applications in mobile ad-hoc networks: Research trends* (pp. 341–372). Hershey, PA: IGI Global. doi:10.4018/978-1-4666-0321-9.ch019.

Ahamed, S. I., Buford, J. F., Sharmin, M., Haque, M. M., & Talukder, N. (2008). Secure service discovery. In Zhang, Y., Zheng, J., & Ma, M. (Eds.), *Handbook of research on wireless security* (pp. 11–27). Hershey, PA: IGI Global. doi:10.4018/978-1-59904-899-4.ch002.

Ahmed, A. (2009). Employee surveillance based on free text detection of keystroke dynamics. In Gupta, M., & Sharman, R. (Eds.), *Handbook of research on social and organizational liabilities in information security* (pp. 47–63). Hershey, PA: IGI Global.

Ahmed, A. A., & Traoré, I. (2012). Performance metrics and models for continuous authentication systems. In Traore, I., & Ahmed, A. (Eds.), *Continuous authentication using biometrics: Data, models, and metrics* (pp. 23–39). Hershey, PA: IGI Global.

Aikins, S. K. (2010). E-Planning: Information security risks and management implications. In Silva, C. (Ed.), *Handbook of research on e-planning: ICTs for urban development and monitoring* (pp. 404–419). Hershey, PA: IGI Global. doi:10.4018/978-1-61520-929-3.ch021.

Akkaladevi, S., Katangur, A. K., & Luo, X. (2009). Protein structure prediction by fusion, bayesian methods. In Rabuñal Dopico, J., Dorado, J., & Pazos, A. (Eds.), *Encyclopedia of artificial intelligence* (pp. 1330–1336). Hershey, PA: IGI Global.

Albuquerque, S. L., & Gondim, P. R. (2012). Applying continuous authentication to protect electronic transactions. In Chou, T. (Ed.), *Information assurance and security technologies for risk assessment and threat management: Advances* (pp. 134–161). Hershey, PA: IGI Global.

Aldas-Manzano, J., Ruiz-Mafe, C., & Sanz-Blas, S. (2009). Mobile commerce adoption in Spain: The influence of consumer attitudes and ICT usage behaviour. In Unhelkar, B. (Ed.), *Handbook of research in mobile business: Technical, methodological and social perspectives* (2nd ed., pp. 282–292). Hershey, PA: IGI Global.

Aleksic, P. S., & Katsaggelos, A. K. (2009). Lip feature extraction and feature evaluation in the context of speech and speaker recognition. In Liew, A., & Wang, S. (Eds.), *Visual speech recognition: Lip segmentation and mapping* (pp. 39–69). Hershey, PA: IGI Global. doi:10.4018/978-1-60566-186-5.ch002.

Alfimtsev, A., Sakulin, S., & Devyatkov, V. (2012). Web personalization based on fuzzy aggregation and recognition of user activity. [IJWP]. *International Journal of Web Portals*, *4*(1), 33–41. doi:10.4018/jwp.2012010103.

Alonso, J. M., Castiello, C., Lucarelli, M., & Mencar, C. (2012). Modeling interpretable fuzzy rule-based classifiers for medical decision support. In Magdalena-Benedito, R., Soria-Olivas, E., Martínez, J., Gómez-Sanchis, J., & Serrano-López, A. (Eds.), *Medical applications of intelligent data analysis: Research advancements* (pp. 255–272). Hershey, PA: IGI Global. doi:10.4018/978-1-4666-1803-9.ch017.

Amin, M. A., & Yan, H. (2010). Gabor wavelets in behavioral biometrics. In Wang, L., & Geng, X. (Eds.), *Behavioral biometrics for human identification: Intelligent applications* (pp. 121–150). Hershey, PA: IGI Global.

Andò, B., Baglio, S., La Malfa, S., & Marletta, V. (2011). Innovative smart sensing solutions for the visually impaired. In Pereira, J. (Ed.), *Handbook of research on personal autonomy technologies and disability informatics* (pp. 60–74). Hershey, PA: IGI Global.

André, E. (2008). Design and evaluation of embodied conversational agents for educational and advisory software. In Luppicini, R. (Ed.), *Handbook of conversation design for instructional applications* (pp. 343–362). Hershey, PA: IGI Global. doi:10.4018/978-1-59904-597-9.ch020.

Andriole, S. J. (2009). Business technology trends analysis. In S. Andriole (Authored) (Ed.), *Technology due diligence: Best practices for chief information officers, venture capitalists, and technology vendors* (pp. 99-134). Hershey, PA: IGI Global. doi: doi:10.4018/978-1-60566-018-9.ch003.

Ang, L., Lim, K. H., Seng, K. P., & Chin, S. W. (2010). A Lyapunov Theory-based neural network approach for face recognition. In Chiong, R. (Ed.), *Intelligent systems for automated learning and adaptation: Emerging trends and applications* (pp. 23–48). Hershey, PA: IGI Global.

Årnes, A. (2008). Large-scale monitoring of critical digital infrastructures. In Janczewski, L., & Colarik, A. (Eds.), *Cyber warfare and cyber terrorism* (pp. 273–280). Hershey, PA: IGI Global.

Bagnato, A., Kordy, B., Meland, P. H., & Schweitzer, P. (2012). Attribute decoration of attack–defense trees. [IJSSE]. *International Journal of Secure Software Engineering*, *3*(2), 1–35. doi:10.4018/jsse.2012040101.

Bai, Y., & Khan, K. M. (2013). Ell secure information system using modal logic technique. In Khan, K. (Ed.), *Developing and evaluating security-aware software systems* (pp. 125–137). Hershey, PA: IGI Global.

Barnaghi, P. M., Wang, W., & Kurian, J. C. (2009). Semantic association analysis in ontology-based information retrieval. In Theng, Y., Foo, S., Goh, D., & Na, J. (Eds.), *Handbook of research on digital libraries: Design, development, and impact* (pp. 131–141). Hershey, PA: IGI Global. doi:10.4018/978-1-59904-879-6.ch013.

Becker, K., & Parker, J. R. (2009). On choosing games and what counts as a "good" game. In Ferdig, R. (Ed.), *Handbook of research on effective electronic gaming in education* (pp. 636–651). Hershey, PA: IGI Global.

Belblidia, M. S. (2010). Building community resilience through social networking sites: Using online social networks for emergency management. [IJIS-CRAM]. *International Journal of Information Systems for Crisis Response and Management*, 2(1), 24–36. doi:10.4018/jiscrm.2010120403.

Belu, R. (2013). Artificial intelligence techniques for solar energy and photovoltaic applications. In Anwar, S., Efstathiadis, H., & Qazi, S. (Eds.), *Handbook of research on solar energy systems and technologies* (pp. 376–436). Hershey, PA: IGI Global.

Beynon, M. (2008). Fuzzy decision-tree-based analysis of databases. In Galindo, J. (Ed.), *Handbook of research on fuzzy information processing in databases* (pp. 760–783). Hershey, PA: IGI Global. doi:10.4018/978-1-59904-853-6.ch031.

Bhattacharyya, S. (2012). Neural networks: Evolution, topologies, learning algorithms and applications. In Mago, V., & Bhatia, N. (Eds.), *Cross-disciplinary applications of artificial intelligence and pattern recognition: Advancing technologies* (pp. 450–498). Hershey, PA: IGI Global.

Bhatti, M. W., Wang, Y., & Guan, L. (2008). Language independent recognition of human emotion using artificial neural networks. [IJCINI]. *International Journal of Cognitive Informatics and Natural Intelligence*, 2(3), 1–21. doi:10.4018/jcini.2008070101.

Bhowmik, M. K., Saha, P., Majumder, G., & Bhattacharjee, D. (2013). Decision fusion of multisensor images for human face identification in information security. In Bhattacharyya, S., & Dutta, P. (Eds.), *Handbook of research on computational intelligence for engineering, science, and business* (pp. 571–591). Hershey, PA: IGI Global.

Bin, M., Chun-lei, L., Yun-hong, W., & Xiao, B. (2011). Salient region detection for biometric watermarking. In Wang, J., Cheng, J., & Jiang, S. (Eds.), *Computer vision for multimedia applications: methods and solutions* (pp. 218–236). Hershey, PA: IGI Global.

Bisdikian, C., Kaplan, L. M., Srivastava, M. B., Thornley, D. J., Verma, D., & Young, R. I. (2010). Quality of sensor-originated information in coalition information networks. In Verma, D. (Ed.), *Network science for military coalition operations: information exchange and interaction* (pp. 15–41). Hershey, PA: IGI Global. doi:10.4018/978-1-61520-855-5.ch002.

Biswas, J., Tolstikov, A., Aung, A., Foo, V. S., & Huang, W. (2011). Ambient intelligence for elder-care – The nuts and bolts: Sensor data acquisition, processing and activity recognition under resource constraints. In Chong, N., & Mastrogiovanni, F. (Eds.), *Handbook of research on ambient intelligence and smart environments: Trends and perspectives* (pp. 392–423). Hershey, PA: IGI Global. doi:10.4018/978-1-61692-857-5.ch020.

Blažic, S., Guechi, E., Lauber, J., Dambrine, M., & Klancar, G. (2010). Path planning and path tracking of industrial mobile robots. In Rigatos, G. (Ed.), *Intelligent industrial systems: Modeling, automation and adaptive behavior* (pp. 84–124). Hershey, PA: IGI Global. doi:10.4018/978-1-61520-849-4.ch004.

Blind, K., & Gauch, S. (2009). The demand for e-government standards. In Jakobs, K. (Ed.), *Information communication technology standardization for e-business sectors: Integrating Supply and demand factors* (pp. 9–23). Hershey, PA: IGI Global. doi:10.4018/978-1-60566-320-3.ch002.

Boloori, A., & Mahmoudi, M. (2013). Networks flow applications. In Farahani, R., & Miandoabchi, E. (Eds.), *Graph theory for operations research and management: Applications in industrial engineering* (pp. 246–256). Hershey, PA: IGI Global.

Bountis, C. (2009). Distributed knowledge management in healthcare. In Lazakidou, A., & Siassiakos, K. (Eds.), *Handbook of research on distributed medical informatics and e-health* (pp. 198–214). Hershey, PA: IGI Global.

Brodsky, J., & Radvanovsky, R. (2011). Control systems security. In Holt, T., & Schell, B. (Eds.), *Corporate hacking and technology-driven crime: Social dynamics and implications* (pp. 187–204). Hershey, PA: IGI Global.

Brooks, D. (2013). Security threats and risks of intelligent building systems: Protecting facilities from current and emerging vulnerabilities. In Laing, C., Badii, A., & Vickers, P. (Eds.), *Securing critical infrastructures and critical control systems: Approaches for threat protection* (pp. 1–16). Hershey, PA: IGI Global.

Brosso, I., & La Neve, A. (2012). Information security management based on adaptive security policy using user behavior analysis. In Gupta, M., Walp, J., & Sharman, R. (Eds.), *Strategic and practical approaches for information security governance: Technologies and applied solutions* (pp. 326–345). Hershey, PA: IGI Global. doi:10.4018/978-1-4666-0197-0.ch019.

Byrne, D., Kelly, L., & Jones, G. J. (2012). Multiple multimodal mobile devices: Lessons learned from engineering lifelog solutions. In Alencar, P., & Cowan, D. (Eds.), *Handbook of research on mobile software engineering: Design, implementation, and emergent applications* (pp. 706–724). Hershey, PA: IGI Global. doi:10.4018/978-1-61520-655-1.ch038.

Calandriello, G., & Lioy, A. (2009). Efficient and reliable pseudonymous authentication. In Guo, H. (Ed.), *Automotive informatics and communicative systems: Principles in vehicular networks and data exchange* (pp. 247–263). Hershey, PA: IGI Global. doi:10.4018/978-1-60566-338-8.ch013.

Calderón, C. (2012). A methodology for interactive architecture. In Gu, N., & Wang, X. (Eds.), *Computational design methods and technologies: Applications in CAD, CAM and CAE education* (pp. 274–298). Hershey, PA: IGI Global. doi:10.4018/978-1-61350-180-1.ch016.

Campisi, P., Maiorana, E., & Neri, A. (2010). Privacy enhancing technologies in biometrics. In Li, C. (Ed.), *Handbook of research on computational forensics, digital crime, and investigation: Methods and solutions* (pp. 1–22). Hershey, PA: IGI Global.

Carstens, D. S. (2009). Human and social aspects of password authentication. In Gupta, M., & Sharman, R. (Eds.), *Social and human elements of information security: Emerging trends and countermeasures* (pp. 1–14). Hershey, PA: IGI Global.

Çetingül, H. E., Erzin, E., Yemez, Y., & Tekalp, A. M. (2009). Multimodal speaker identification using discriminative lip motion features. In Liew, A., & Wang, S. (Eds.), *Visual speech recognition: Lip segmentation and mapping* (pp. 463–494). Hershey, PA: IGI Global. doi:10.4018/978-1-60566-186-5.ch016.

Chai, Y. (2010). Computational intelligence-revisited. In Wang, L., & Hong, T. (Eds.), *Intelligent soft computation and evolving data mining: Integrating advanced technologies* (pp. 85–99). Hershey, PA: IGI Global. doi:10.4018/978-1-61520-757-2.ch005.

Chakraborthy, R., Rengamani, H., Kumaraguru, P., & Rao, R. (2011). The UID project: Lessons learned from the West and challenges identified for India. In Santanam, R., Sethumadhavan, M., & Virendra, M. (Eds.), *Cyber security, cyber crime and cyber forensics: Applications and Perspectives* (pp. 1–23). Hershey, PA: IGI Global.

Chakravarty, I. (2009). Online signature recognition. In Wang, J. (Ed.), *Encyclopedia of data warehousing and mining* (2nd ed., pp. 1456–1462). Hershey, PA: IGI Global.

Chandrasekaran, M., & Upadhyaya, S. (2009). A multistage framework to defend against phishing attacks. In Gupta, M., & Sharman, R. (Eds.), *Handbook of research on social and organizational liabilities in information security* (pp. 175–192). Hershey, PA: IGI Global.

Chen, Y., Ding, X., & Wang, P. S. (2009). Dynamic structural statistical model based online signature verification. [IJDCF]. *International Journal of Digital Crime and Forensics, 1*(3), 21–41. doi:10.4018/jdcf.2009070102.

Chen, Y., & Zhang, Y. (2009). Extracting concepts' relations and users' preferences for personalizing query disambiguation. [IJSWIS]. *International Journal on Semantic Web and Information Systems, 5*(1), 65–79. doi:10.4018/jswis.2009010103.

Cherifi, F., Hemery, B., Giot, R., Pasquet, M., & Rosenberger, C. (2010). Performance evaluation of behavioral biometric systems. In Wang, L., & Geng, X. (Eds.), *Behavioral biometrics for human identification: intelligent applications* (pp. 57–74). Hershey, PA: IGI Global.

Chohra, A., Kanaoui, N., Amarger, V., & Madani, K. (2011). Hybrid intelligent diagnosis approach based on neural pattern recognition and fuzzy decision-making. In Jozefczyk, J., & Orski, D. (Eds.), *Knowledge-based intelligent system advancements: systemic and cybernetic approaches* (pp. 372–394). Hershey, PA: IGI Global. doi:10.4018/978-1-60960-818-7.ch307.

Chowdhury, M. U., & Ray, B. R. (2013). Security risks/vulnerability in a RFID system and possible defenses. In Karmakar, N. (Ed.), *Advanced RFID systems, security, and applications* (pp. 1–15). Hershey, PA: IGI Global.

Clodfelter, R. (2011). Point-of-sale technologies at retail stores: What will the future be like? In Pantano, E., & Timmermans, H. (Eds.), *Advanced technologies management for retailing: Frameworks and cases* (pp. 1–25). Hershey, PA: IGI Global. doi:10.4018/978-1-60960-738-8.ch001.

Costa-Montenegro, E., Peleteiro, A. M., Burguillo, J. C., Vales-Alonso, J., & Barragáns-Martínez, A. B. (2013). On the dissemination of IEEE 802.11p Warning messages in distributed vehicular urban networks. In A. Loo (Ed.), Distributed computing innovations for business, engineering, and science (pp. 234-252). Hershey, PA: IGI Global. doi: doi:10.4018/978-1-4666-2533-4.ch012.

Coutinho, M. P., Lambert-Torres, G., Borges da Silva, L. E., Lazarek, H., & Franz, E. (2013). Detecting cyber attacks on SCADA and other critical infrastructures. In Laing, C., Badii, A., & Vickers, P. (Eds.), *Securing critical infrastructures and critical control systems: Approaches for threat protection* (pp. 17–53). Hershey, PA: IGI Global.

Cruz-Cunha, M. M., Putnik, G. D., & Varajão, J. (2012). The use of customer relationship management software in meta-enterprises for virtual enterprise integration. In Colomo-Palacios, R., Varajão, J., & Soto-Acosta, P. (Eds.), *Customer relationship management and the social and semantic web: Enabling Cliens Conexus* (pp. 312–326). Hershey, PA: IGI Global.

Cruz-Cunha, M. M., Simões, R., Tavares, A., & Miranda, I. (2010). GuiMarket: An E-marketplace of healthcare and social care services for individuals with special needs. In Cruz-Cunha, M., Tavares, A., & Simoes, R. (Eds.), *Handbook of research on developments in e-health and telemedicine: Technological and social perspectives* (pp. 904–917). Hershey, PA: IGI Global. doi:10.4018/978-1-61520-967-5.ch072.

Cunha, M. M., Putnik, G. D., da Silva, J. P., & Oliveira Santos, J. P. (2008). Technologies to support the market of resources as an infrastructure for agile/virtual enterprise integration. In Protogeros, N. (Ed.), *Agent and web service technologies in virtual enterprises* (pp. 76–96). Hershey, PA: IGI Global.

Cunha, M. M., Putnik, G. D., & Silva, J. P. (2008). Market of resources: Supporting technologies. In Putnik, G., & Cruz-Cunha, M. (Eds.), *Encyclopedia of networked and virtual organizations* (pp. 906–912). Hershey, PA: IGI Global. doi:10.4018/978-1-59904-885-7.ch119.

Cuzzolin, F. (2010). Multilinear modeling for robust identity recognition from gait. In Wang, L., & Geng, X. (Eds.), *Behavioral biometrics for human identification: Intelligent applications* (pp. 169–188). Hershey, PA: IGI Global.

D'Aubeterre, F., Iyer, L. S., Ehrhardt, R., & Singh, R. (2009). Discovery process in a B2B eMarketplace: A semantic matchmaking approach. [IJIIT]. *International Journal of Intelligent Information Technologies*, 5(4), 16–40. doi:10.4018/jiit.2009080702.

D'Ulizia, A. (2009). Exploring multimodal input fusion strategies. In Grifoni, P. (Ed.), *Multimodal human computer interaction and pervasive services* (pp. 34–57). Hershey, PA: IGI Global. doi:10.4018/978-1-60566-386-9.ch003.

D'Ulizia, A., Ferri, F., Grifoni, P., & Guzzo, T. (2010). Smart homes to support elderly people: Innovative technologies and social impacts. In Coronato, A., & De Pietro, G. (Eds.), *Pervasive and smart technologies for healthcare: Ubiquitous methodologies and tools* (pp. 25–38). Hershey, PA: IGI Global. doi:10.4018/978-1-61520-765-7.ch002.

Das, T. (2009). Movement prediction oriented adaptive location management. In Khalil, I. (Ed.), *Handbook of research on mobile multimedia* (2nd ed., pp. 464–483). Hershey, PA: IGI Global.

De Marsico, M., De Marsico, M., Nappi, M., Nappi, M., Riccio, D., & Riccio, D. et al. (2011). Face searching in large databases. In Zhang, Y. (Ed.), *Advances in face image analysis: Techniques and technologies* (pp. 16–41). Hershey, PA: IGI Global.

194

Deb, S. (2009). Emergence index in image databases. In Khosrow-Pour, M. (Ed.), *Encyclopedia of information science and technology* (2nd ed., pp. 1361–1365). Hershey, PA: IGI Global.

Dehkordi, L. F., Ghorbani, A., & Aliahmadi, A. R. (2011). Application of RFID technology in banking sector. In Sarlak, M., & Hastiani, A. (Eds.), *E-Banking and emerging multidisciplinary processes: Social, economical and organizational models* (pp. 149–163). Hershey, PA: IGI Global.

Derrac, J., García, S., & Herrera, F. (2010). A survey on evolutionary instance selection and generation. [IJAMC]. *International Journal of Applied Metaheuristic Computing, 1*(1), 60–92. doi:10.4018/jamc.2010102604.

Desjardins, C., Laumônier, J., & Chaib-draa, B. (2009). Learning agents for collaborative driving. In Bazzan, A., & Klügl, F. (Eds.), *Multi-agent systems for traffic and transportation engineering* (pp. 240–260). Hershey, PA: IGI Global. doi:10.4018/978-1-60566-226-8.ch011.

Devasenapati, S. B., & Ramachandran, K. I. (2012). Artificial intelligence based green technology retrofit for misfire detection in old engines. [IJGC]. *International Journal of Green Computing, 3*(1), 43–55. doi:10.4018/jgc.2012010104.

Devyatkov, V., & Alfimtsev, A. (2011). Human-computer interaction in games using computer vision techniques. In Cruz-Cunha, M., Varvalho, V., & Tavares, P. (Eds.), *Business, technological, and social dimensions of computer games: Multidisciplinary developments* (pp. 146–167). Hershey, PA: IGI Global. doi:10.4018/978-1-60960-567-4.ch010.

Dhanalakshmi, R., & Chellappan, C. (2012). Fraud and identity theft issues. In Gupta, M., Walp, J., & Sharman, R. (Eds.), *Strategic and practical approaches for information security governance: Technologies and applied solutions* (pp. 245–260). Hershey, PA: IGI Global. doi:10.4018/978-1-4666-0197-0.ch014.

Doloc-Mihu, A. (2009). Modeling score distributions. In Wang, J. (Ed.), *Encyclopedia of data warehousing and mining* (2nd ed., pp. 1330–1336). Hershey, PA: IGI Global.

dos Santos Silva, M. P., Câmara, G., & Escada, M. I. (2009). Image mining: Detecting deforestation patterns through satellites. In Rahman, H. (Ed.), *Data mining applications for empowering knowledge societies* (pp. 55–75). Hershey, PA: IGI Global.

Doyle, D. J. (2009). Medical privacy and the internet. In Lazakidou, A., & Siassiakos, K. (Eds.), *Handbook of research on distributed medical informatics and e-health* (pp. 17–29). Hershey, PA: IGI Global.

Du, Y. E. (2009). Biometric technologies. In Khosrow-Pour, M. (Ed.), *Encyclopedia of information science and technology* (2nd ed., pp. 369–374). Hershey, PA: IGI Global.

Durga, S. (2010). A progressive exposure approach for secure service discovery in pervasive computing environments. In Godara, V. (Ed.), *Strategic pervasive computing applications: Emerging trends* (pp. 111–122). Hershey, PA: IGI Global.

Dwivedi, Y. K. (2009). A bibliometric analysis of electronic government research. In Sahu, G., Dwivedi, Y., & Weerakkody, V. (Eds.), *E-Government development and diffusion: Inhibitors and facilitators of digital democracy* (pp. 176–256). Hershey, PA: IGI Global. doi:10.4018/978-1-60566-713-3.ch012.

El-Darymli, K., & Ahmed, M. H. (2012). Wireless sensor network testbeds: A survey. In Zaman, N., Ragab, K., & Abdullah, A. (Eds.), *Wireless sensor networks and energy efficiency: Protocols, routing and management* (pp. 148–205). Hershey, PA: IGI Global. doi:10.4018/978-1-4666-0101-7.ch007.

Erlich, Z., & Zviran, M. (2012). Goals and practices in maintaining information systems security. In Nemati, H. (Ed.), *Optimizing information security and advancing privacy assurance: New technologies* (pp. 214–224). Hershey, PA: IGI Global.

Espina, J., Baldus, H., Falck, T., Garcia, O., & Klabunde, K. (2009). Towards easy-to-use, safe, and secure wireless medical body sensor networks. In Olla, P., & Tan, J. (Eds.), *Mobile health solutions for biomedical applications* (pp. 159–179). Hershey, PA: IGI Global. doi:10.4018/978-1-60566-332-6.ch009.

Faraj, M. I., & Bigun, J. (2009). Lip motion features for biometric person recognition. In Liew, A., & Wang, S. (Eds.), *Visual speech recognition: Lip segmentation and mapping* (pp. 495–532). Hershey, PA: IGI Global. doi:10.4018/978-1-60566-186-5.ch017.

Farzaneh, M., Vanani, I. R., & Sohrabi, B. (2012). Utilization of intelligent software agent features for improving e-learning efforts: A comprehensive investigation. [IJVPLE]. *International Journal of Virtual and Personal Learning Environments, 3*(1), 55–68. doi: doi:10.4018/IJVPLE.2012010104.

Fernández, D. R. (2010). An agent-based architecture to ubiquitous health. In Mohammed, S., & Fiaidhi, J. (Eds.), *Ubiquitous health and medical informatics: The ubiquity 2.0 trend and beyond* (pp. 213–232). Hershey, PA: IGI Global. doi:10.4018/978-1-61520-777-0.ch011.

Fiedrich, F., & Van de Walle, B. (2009). The fifth international conference on information systems for crisis response and management, Washington DC, May 4-7 2008. [IJISCRAM]. *International Journal of Information Systems for Crisis Response and Management, 1*(1), 70–74. doi:10.4018/jiscrm.2009010106.

Francisco Vargas, J., & Ferrer, M. A. (2009). Neural networks on handwritten signature verification. In Rabuñal Dopico, J., Dorado, J., & Pazos, A. (Eds.), *Encyclopedia of artificial intelligence* (pp. 1232–1237). Hershey, PA: IGI Global.

Fung, M. Y., & Paynter, J. (2008). The impact of information technology in healthcare privacy. In Duquenoy, P., George, C., & Kimppa, K. (Eds.), *Ethical, legal and social issues in medical informatics* (pp. 186–227). Hershey, PA: IGI Global. doi:10.4018/978-1-59904-780-5.ch009.

Ganjigatti, J. P., & Pratihar, D. K. (2009). Design and development of knowledge bases for forward and reverse mappings of TIG welding process. In Wang, H. (Ed.), *Intelligent data analysis: Developing new methodologies through pattern discovery and recovery* (pp. 185–200). Hershey, PA: IGI Global.

Gavrilova, M. L. (2012). Adaptive computation paradigm in knowledge representation: Traditional and emerging applications. In Wang, Y. (Ed.), *Software and intelligent sciences: New transdisciplinary findings* (pp. 142–156). Hershey, PA: IGI Global. doi:10.4018/978-1-4666-0261-8.ch009.

Ghosh, S. (2010). Net-centric military to civilian transformation. In Ghosh, S. (Ed.), *Net centricity and technological interoperability in organizations: Perspectives and strategies* (pp. 84–98). Hershey, PA: IGI Global.

Gill, S., & Paranjape, R. (2010). A review of recent contribution in agent-based health care modeling. In Paranjape, R., & Sadanand, A. (Eds.), *Multi-agent systems for healthcare simulation and modeling: Applications for system improvement* (pp. 26–43). Hershey, PA: IGI Global.

196

Gill, S. K., & Cormican, K. (2008). Ambient intelligent (AmI) systems development. In Zhao, F. (Ed.), *Information technology entrepreneurship and innovation* (pp. 1–22). Hershey, PA: IGI Global. doi:10.4018/978-1-59904-901-4.ch001.

Gilles, R., James, T., Barkhi, R., & Diamantaras, D. (2009). Simulating social network formation: A case-based decision theoretic model. [IJVCSN]. *International Journal of Virtual Communities and Social Networking*, *1*(4), 1–20. doi:10.4018/jvcsn.2009092201.

Gingras, D. (2009). An overview of positioning and data fusion techniques applied to land vehicle navigation systems. In Guo, H. (Ed.), *Automotive informatics and communicative systems: Principles in vehicular networks and data exchange* (pp. 219–246). Hershey, PA: IGI Global. doi:10.4018/978-1-60566-338-8.ch012.

Glancy, F. H., & Yadav, S. B. (2011). Business intelligence conceptual model. [IJBIR]. *International Journal of Business Intelligence Research*, *2*(2), 48–66. doi:10.4018/jbir.2011040104.

Golfarelli, M., Mandreoli, F., Penzo, W., Rizzi, S., & Turricchia, E. (2012). BIN: Business intelligence networks. In Zorrilla, M., Mazón, J., Ferrández, Ó., Garrigós, I., Daniel, F., & Trujillo, J. (Eds.), *Business intelligence applications and the web: Models, systems and technologies* (pp. 244–265). Hershey, PA: IGI Global.

Gomez-Skarmeta, A., Mendez, A. P., Garcia, E. T., & Millán, G. L. (2012). User-centric privacy management in future network infrastructure. In Yee, G. (Ed.), *Privacy protection measures and technologies in business organizations: Aspects and standards* (pp. 32–64). Hershey, PA: IGI Global.

Goteng, G., Tiwari, A., & Roy, R. (2009). Grid computing: Combating global terrorism with the world wide grid. In Udoh, E., & Wang, F. (Eds.), *Handbook of research on grid technologies and utility computing: Concepts for managing large-scale applications* (pp. 269–280). Hershey, PA: IGI Global. doi:10.4018/978-1-60566-184-1.ch027.

Grandison, T., Hsueh, P. S., Zeng, L., Chang, H., Chen, Y., & Lan, C. et al. (2012). Privacy protection issues for healthcare wellness clouds. In Yee, G. (Ed.), *Privacy protection measures and technologies in business organizations: Aspects and standards* (pp. 227–244). Hershey, PA: IGI Global.

Guan, S. (2008). Integrity protection of mobile agent data. In Garson, G., & Khosrow-Pour, M. (Eds.), *Handbook of research on public information technology* (pp. 423–462). Hershey, PA: IGI Global. doi:10.4018/978-1-59904-857-4.ch043.

Guerrero-Ibáñez, J., Flores-Cortés, C., & Damián-Reyes, P. (2012). Development of applications for vehicular communication network environments: Challenges and opportunities. In Aquino-Santos, R., Edwards, A., & Rangel-Licea, V. (Eds.), *Wireless technologies in vehicular ad hoc networks: Present and future challenges* (pp. 183–204). Hershey, PA: IGI Global. doi:10.4018/978-1-4666-0209-0.ch009.

Guest, C. L., & Guest, J. M. (2011). Legal issues in the use of technology in higher education: Copyright and privacy in the academy. In Surry, D., Gray, R. Jr, & Stefurak, J. (Eds.), *Technology integration in higher education: Social and organizational aspects* (pp. 72–85). Hershey, PA: IGI Global.

Gumzej, R., & Lipicnik, M. (2010). Information and communication technology in logistics as a comparative advantage. In Luo, Z. (Ed.), *Service science and logistics informatics: Innovative perspectives* (pp. 144–156). Hershey, PA: IGI Global. doi:10.4018/978-1-61520-603-2.ch008.

Gupta, C. N., & Palaniappan, R. (2009). Biometric paradigm using visual evoked potential. In Khosrow-Pour, M. (Ed.), *Encyclopedia of information science and technology* (2nd ed., pp. 362–368). Hershey, PA: IGI Global.

Gupta, M., Rao, R., & Upadhyaya, S. (2008). Electronic banking and information assurance issues: Survey and synthesis. In Ravi, V. (Ed.), *Advances in banking technology and management: Impacts of ICT and CRM* (pp. 119–138). Hershey, PA: IGI Global.

Gurau, C. (2009). How good is your shopping agent? Users' perception regarding shopping agents' service quality. In Wan, Y. (Ed.), *Comparison-shopping services and agent designs* (pp. 151–164). Hershey, PA: IGI Global. doi:10.4018/978-1-59904-978-6.ch010.

Guru, D. S., Nagasundara, K. B., Manjunath, S., & Dinesh, R. (2011). An approach for hand vein representation and indexing. [IJDCF]. *International Journal of Digital Crime and Forensics*, 3(2), 1–15. doi:10.4018/jdcf.2011040101.

Ha, H., & Coghill, K. (2009). Current measures to protect e-consumers' privacy in australia. In Chen, K., & Fadlalla, A. (Eds.), *Online consumer protection: Theories of human relativism* (pp. 123–150). Hershey, PA: IGI Global.

Habib, M. K. (2011). Humanitarian demining action plan: Humanity and technological challenges. In Vargas Martin, M., Garcia-Ruiz, M., & Edwards, A. (Eds.), *Technology for facilitating humanity and combating social deviations: interdisciplinary perspectives* (pp. 114–131). Hershey, PA: IGI Global.

Hájek, P., & Olej, V. (2011). Air quality assessment by neural networks. In Olej, V., Obršálová, I., & Krupka, J. (Eds.), *Environmental modeling for sustainable regional development: System approaches and advanced methods* (pp. 91–117). Hershey, PA: IGI Global.

Hameed, W. W. (2011). A secure electronic voting. In Al Ajeeli, A., & Al-Bastaki, Y. (Eds.), *Handbook of research on e-services in the public sector: E-Government strategies and advancements* (pp. 431–449). Hershey, PA: IGI Global.

Hamidi, H. (2009). Modeling fault tolerant and secure mobile agent execution in distributed systems. In Sugumaran, V. (Ed.), *Distributed artificial intelligence, agent technology, and collaborative applications* (pp. 132–146). Hershey, PA: IGI Global.

Han, D. C., & Barber, S. K. (2012). Simulating UAV surveillance for analyzing impact of commitments in multi-agent systems. [IJATS]. *International Journal of Agent Technologies and Systems*, 4(1), 1–16. doi:10.4018/jats.2012010101.

Haque, E., & Yoshida, N. (2012). Clustering in wireless sensor networks: Context-aware approaches. In Abawajy, J., Pathan, M., Rahman, M., Pathan, A., & Deris, M. (Eds.), *Internet and distributed computing advancements: Theoretical frameworks and practical applications* (pp. 197–211). Hershey, PA: IGI Global. doi:10.4018/978-1-4666-0161-1.ch008.

Harnett, B. (2011). Patient centered medicine and technology adaptation. In Röcker, C., & Ziefle, M. (Eds.), *E-Health, assistive technologies and applications for assisted living: Challenges and solutions* (pp. 1–22). Hershey, PA: IGI Global. doi:10.4018/978-1-60960-469-1.ch001.

Henniger, O. (2010). Security evaluation of behavioral biometric systems. In Wang, L., & Geng, X. (Eds.), *Behavioral biometrics for human identification: intelligent applications* (pp. 44–56). Hershey, PA: IGI Global.

198

Hiratsuka, M., Ito, K., Aoki, T., & Higuchi, T. (2009). Toward biomolecular computers using reaction-diffusion dynamics. [IJNMC]. *International Journal of Nanotechnology and Molecular Computation*, *1*(3), 17–25. doi:10.4018/jnmc.2009070102.

Huang, C., & Lin, H. (2011). Patent infringement risk analysis using rough set theory. In Zhang, Q., Segall, R., & Cao, M. (Eds.), *Visual analytics and interactive technologies: Data, text and web mining applications* (pp. 123–150). Hershey, PA: IGI Global.

Hughes, J., & Robinson, S. (2009). Social structures of online religious communities. In Hatzipanagos, S., & Warburton, S. (Eds.), *Handbook of research on social software and developing community ontologies* (pp. 193–207). Hershey, PA: IGI Global. doi:10.4018/978-1-60566-208-4.ch014.

Iglezakis, I. (2009). Protecting identity without comprising privacy: Privacy implications of identity protection. In Politis, D., Kozyris, P., & Iglezakis, I. (Eds.), *Socioeconomic and legal implications of electronic intrusion* (pp. 62–88). Hershey, PA: IGI Global. doi:10.4018/978-1-60566-204-6.ch004.

Im, S., & Ras, Z. W. (2009). Data confidentiality and chase-based knowledge discovery. In Wang, J. (Ed.), *Encyclopedia of data warehousing and mining* (2nd ed., pp. 361–366). Hershey, PA: IGI Global.

Ionescu, B., Marin, A., Lambert, P., Coquin, D., & Vertan, C. (2010). A content-driven system architecture for tackling automatic cataloging of animated movie databases. [IJDLS]. *International Journal of Digital Library Systems*, *1*(2), 1–23. doi:10.4018/jdls.2010040101.

Jacobsen, R. H., Toftegaard, T. S., & Kjærgaard, J. K. (2012). IP connected low power wireless personal area networks in the future internet. In Prakash Vidyarthi, D. (Ed.), *Technologies and protocols for the future of internet design: Reinventing the web* (pp. 191–213). Hershey, PA: IGI Global. doi:10.4018/978-1-4666-0203-8.ch010.

Jadhav, D. V., Dattatray, V. J., Raghunath, S. H., & Holambe, R. S. (2011). Transform based feature extraction and dimensionality reduction techniques. In Zhang, Y. (Ed.), *Advances in face image analysis: Techniques and technologies* (pp. 120–136). Hershey, PA: IGI Global.

James, F., & Gurram, R. (2008). Multimodal and federated interaction. In Mühlhäuser, M., & Gurevych, I. (Eds.), *Handbook of research on ubiquitous computing technology for real time enterprises* (pp. 487–507). Hershey, PA: IGI Global. doi:10.4018/978-1-59904-832-1.ch021.

Jansen, T. W. (2009). Practical privacy assessments. In Chen, K., & Fadlalla, A. (Eds.), *Online consumer protection: Theories of human relativism* (pp. 57–84). Hershey, PA: IGI Global.

Jasemian, Y. (2009). Patient monitoring in diverse environments. In Olla, P., & Tan, J. (Eds.), *Mobile health solutions for biomedical applications* (pp. 129–142). Hershey, PA: IGI Global. doi:10.4018/978-1-60566-332-6.ch007.

Jerbi, M., Senouci, S., Ghamri-Doudane, Y., & Cherif, M. (2010). Vehicular communications networks: Current trends and challenges. In Pierre, S. (Ed.), *Next generation mobile networks and ubiquitous computing* (pp. 251–262). Hershey, PA: IGI Global. doi:10.4018/978-1-60566-250-3.ch023.

Jiao, Y., Hurson, A. R., & Potok, T. E. (2009). Mobile agent-based information systems and security. In Khosrow-Pour, M. (Ed.), *Encyclopedia of information science and technology* (2nd ed., pp. 2574–2579). Hershey, PA: IGI Global.

Jones, J. D. (2008). Logic programming languages for expert systems. In Adam, F., & Humphreys, P. (Eds.), *Encyclopedia of decision making and decision support technologies* (pp. 593–603). Hershey, PA: IGI Global. doi:10.4018/978-1-59904-843-7.ch066.

Just, M., & Renaud, K. (2012). Trends in government e-authentication: Policy and practice. In Bwalya, K., & Zulu, S. (Eds.), *Handbook of research on e-government in emerging economies: Adoption, e-participation, and legal frameworks* (pp. 664–677). Hershey, PA: IGI Global. doi:10.4018/978-1-4666-0324-0.ch034.

Kadaba, R., Budalakoti, S., DeAngelis, D., & Barber, K. S. (2011). Modeling virtual footprints. [IJATS]. *International Journal of Agent Technologies and Systems*, *3*(2), 1–17. doi:10.4018/jats.2011040101.

Kahraman, C., Çebi, S., & Kaya, I. (2011). Group decision making for advanced manufacturing technology selection using the choquet integral. In Yearwood, J., & Stranieri, A. (Eds.), *Technologies for supporting reasoning communities and collaborative decision making: Cooperative approaches* (pp. 193–212). Hershey, PA: IGI Global.

Kakabadse, N. K., Kouzmin, A., & Kakabadse, A. (2012). Radio-frequency identification and human tagging: Newer coercions. In Romm Livermore, C. (Ed.), *E-Politics and organizational implications of the internet: Power, influence, and social change* (pp. 1–18). Hershey, PA: IGI Global. doi:10.4018/978-1-4666-0966-2.ch001.

Kalogirou, S., Metaxiotis, K., & Mellit, A. (2010). Artificial intelligence techniques for modern energy applications. In Metaxiotis, K. (Ed.), *Intelligent information systems and knowledge management for energy: Applications for decision support, usage, and environmental protection* (pp. 1–39). Hershey, PA: IGI Global.

Kanellopoulos, D. (2009). Intelligent technologies for tourism. In Khosrow-Pour, M. (Ed.), *Encyclopedia of information science and technology* (2nd ed., pp. 2141–2146). Hershey, PA: IGI Global.

Karantjias, A., & Polemi, N. (2012). Effective guidelines for facilitating construction of successful, advanced, user-centric IAM frameworks. In Sharman, R., Das Smith, S., & Gupta, M. (Eds.), *Digital identity and access management: Technologies and frameworks* (pp. 39–63). Hershey, PA: IGI Global.

Karastoyanova, D., van Lessen, T., Leymann, F., Ma, Z., Nitzche, J., & Wetzstein, B. (2009). Semantic business process management: Applying ontologies in BPM. In Cardoso, J., & van der Aalst, W. (Eds.), *Handbook of research on business process modeling* (pp. 299–317). Hershey, PA: IGI Global. doi:10.4018/978-1-60566-288-6.ch014.

Karat, J., Sieck, W., Norman, T. J., Karat, C., Brodie, C., Rasmussen, L., & Sycara, K. (2010). A model for culturally adaptive policy management in ad hoc collaborative contexts. In Verma, D. (Ed.), *Network science for military coalition operations: Information exchange and interaction* (pp. 174–190). Hershey, PA: IGI Global. doi:10.4018/978-1-61520-855-5.ch009.

Keng Ang, K., & Quek, C. (2009). Rough set-based neuro-fuzzy system. In Rabuñal Dopico, J., Dorado, J., & Pazos, A. (Eds.), *Encyclopedia of artificial intelligence* (pp. 1396–1403). Hershey, PA: IGI Global.

Kern, J., Fister, K., & Polasek, O. (2009). Active patient role in recording health data. In Khosrow-Pour, M. (Ed.), *Encyclopedia of information science and technology* (2nd ed., pp. 14–19). Hershey, PA: IGI Global. doi:10.4018/978-1-60566-988-5.ch099.

Khaled, Y., Tsukada, M., Santa, J., & Ernst, T. (2010). The role of communication technologies in vehicular applications. In Watfa, M. (Ed.), *Advances in vehicular ad-hoc networks: developments and challenges* (pp. 37–58). Hershey, PA: IGI Global. doi:10.4018/978-1-61520-913-2.ch003.

Khemakhem, S., Drira, K., & Jmaiel, M. (2011). Description, classification and discovery approaches for software components: A comparative study. In Dogru, A., & Biçer, V. (Eds.), *Modern software engineering concepts and practices: Advanced approaches* (pp. 196–219). Hershey, PA: IGI Global.

Khosla, M., Sarin, R. K., Uddin, M., Singh, S., & Khosla, A. (2012). Realizing interval type-2 fuzzy systems with type-1 fuzzy systems. In Mago, V., & Bhatia, N. (Eds.), *Cross-disciplinary applications of artificial intelligence and pattern recognition: Advancing technologies* (pp. 412–427). Hershey, PA: IGI Global.

Kim, T. J. (2008). Planning for knowledge cities in ubiquitous technology spaces: Opportunities and challenges. In Yigitcanlar, T., Velibeyoglu, K., & Baum, S. (Eds.), *Creative urban regions: Harnessing urban technologies to support knowledge city initiatives* (pp. 218–230). Hershey, PA: IGI Global. doi:10.4018/978-1-59904-838-3.ch013.

Kirwan, G., & Power, A. (2012). Identity theft and online fraud: What makes us vulnerable to scam artists online? In Kirwan, G., & Power, A. (Eds.), *The psychology of cyber crime: Concepts and principles* (pp. 94–112). Hershey, PA: IGI Global.

Kisku, D. R., Gupta, P., & Sing, J. K. (2011). Graphs in biometrics. In Al-Mutairi, M., & Mohammed, L. (Eds.), *Cases on ICT utilization, practice and solutions: Tools for managing day-to-day issues* (pp. 151–180). Hershey, PA: IGI Global.

Kisku, D. R., Gupta, P., Sing, J. K., Tistarelli, M., & Hwang, C. J. (2012). Low level multispectral palmprint image fusion for large scale biometrics authentication. In Traore, I., & Ahmed, A. (Eds.), *Continuous authentication using biometrics: Data, models, and metrics* (pp. 89–104). Hershey, PA: IGI Global.

Kizza, J., & Migga Kizza, F. (2008). Biometrics for access control. In Kizza, J., & Migga Kizza, F. (Eds.), *Securing the information infrastructure* (pp. 280–296). Hershey, PA: IGI Global.

Kizza, J., & Migga Kizza, F. (2008). Security, anonymity, and privacy. In Kizza, J., & Migga Kizza, F. (Eds.), *Securing the information infrastructure* (pp. 41–64). Hershey, PA: IGI Global.

Klaus, T., & Changchit, C. (2009). Online or traditional: A study to examine course characteristics contributing to students' preference for classroom settings. [IJICTE]. *International Journal of Information and Communication Technology Education*, *5*(3), 14–23. doi:10.4018/jicte.2009070102.

Kljajic, M., & Farr, J. V. (2008). The role of systems engineering in the development of information systems. [IJITSA]. *International Journal of Information Technologies and Systems Approach*, *1*(1), 49–61. doi:10.4018/jitsa.2008010104.

Konstantinidis, K., Sirakoulis, G. C., & Andreadis, I. (2009). Ant colony optimization for use in content based image retrieval. In Mo, H. (Ed.), *Handbook of research on artificial immune systems and natural computing: Applying complex adaptive technologies* (pp. 384–404). Hershey, PA: IGI Global. doi:10.4018/978-1-60566-310-4.ch018.

Kotropoulos, C., & Pitas, I. (2009). Visual speech processing and recognition. In Liew, A., & Wang, S. (Eds.), *Visual speech recognition: Lip segmentation and mapping* (pp. 261–293). Hershey, PA: IGI Global. doi:10.4018/978-1-60566-186-5.ch009.

Krpan, D., Tomaš, S., & Vladušic, R. (2010). Using effect size for group modeling in e-learning systems. In Stankov, S., Glavinic, V., & Rosic, M. (Eds.), *Intelligent tutoring systems in e-learning environments: Design, implementation and evaluation* (pp. 237–257). Hershey, PA: IGI Global. doi:10.4018/978-1-61692-008-1.ch012.

Kuehler, M., Schimke, N., & Hale, J. (2012). Privacy considerations for electronic health records. In Yee, G. (Ed.), *Privacy protection measures and technologies in business organizations: Aspects and standards* (pp. 210–226). Hershey, PA: IGI Global.

Kulkarni, S. (2012). Machine learning approach for content based image retrieval. In Kulkarni, S. (Ed.), *Machine learning algorithms for problem solving in computational applications: Intelligent techniques* (pp. 1–11). Hershey, PA: IGI Global. doi:10.4018/978-1-4666-1833-6.ch001.

Kwak, N. K., & Won Lee, C. (2009). An application of multi-criteria decision-making model to strategic outsourcing for effective supply-chain linkages. In Hunter, M. (Ed.), *Selected readings on strategic information systems* (pp. 223–236). Hershey, PA: IGI Global. doi:10.4018/978-1-60566-677-8.ch092.

Kwiatkowska, M., Atkins, M. S., Matthews, L., Ayas, N. T., & Ryan, C. F. (2009). Knowledge-based induction of clinical prediction rules. In Berka, P., Rauch, J., & Zighed, D. (Eds.), *Data mining and medical knowledge management: Cases and applications* (pp. 350–375). Hershey, PA: IGI Global. doi:10.4018/978-1-60566-218-3.ch017.

Ladner, R. (2009). Electronic government: Overview and issues for national security interests. In Khosrow-Pour, M. (Ed.), *E-Government diffusion, policy, and impact: Advanced issues and practices* (pp. 13–27). Hershey, PA: IGI Global.

Ladner, R., Petry, F., & McCreedy, F. (2010). E-government capabilities for 21st century security and defense. In Weerakkody, V. (Ed.), *Social and organizational developments through emerging e-government applications: New principles and concepts* (pp. 1–13). Hershey, PA: IGI Global.

Lancaster, S. (2009). Biometric controls and privacy. In Chen, K., & Fadlalla, A. (Eds.), *Online consumer protection: Theories of human relativism* (pp. 300–309). Hershey, PA: IGI Global.

Lazarevic, A. (2009). Data mining in security applications. In Wang, J. (Ed.), *Encyclopedia of data warehousing and mining* (2nd ed., pp. 479–485). Hershey, PA: IGI Global.

Lee, S. H., Han, J. H., Leem, Y. T., & Yigitcanlar, T. (2008). Towards ubiquitous city: concept, planning, and experiences in the Republic of Korea. In Yigitcanlar, T., Velibeyoglu, K., & Baum, S. (Eds.), *Knowledge-based urban development: Planning and applications in the information era* (pp. 148–170). Hershey, PA: IGI Global. doi:10.4018/978-1-59904-720-1.ch009.

Lee, S. H., Yigitcanlar, T., & Wong, J. K. (2010). Ubiquitous and smart system approaches to infrastructure planning: Learnings from Korea, Japan and Hong Kong. In Yigitcanlar, T. (Ed.), *Sustainable urban and regional infrastructure development: Technologies, applications and management* (pp. 165–182). Hershey, PA: IGI Global. doi:10.4018/978-1-61520-775-6.ch012.

Leszczyna, R., & Egozcue, E. (2013). ENISA study: Challenges in securing industrial control systems. In Laing, C., Badii, A., & Vickers, P. (Eds.), *Securing critical infrastructures and critical control systems: Approaches for threat protection* (pp. 105–143). Hershey, PA: IGI Global.

Li, J., & Shaw, M. J. (2008). Electronic medical records, HIPAA, and patient privacy. [IJISP]. *International Journal of Information Security and Privacy, 2*(3), 45–54. doi:10.4018/jisp.2008070104.

Li, J., & Shaw, M. J. (2011). Safeguarding the privacy of electronic medical records. In Nemati, H. (Ed.), *Pervasive information security and privacy developments: Trends and advancements* (pp. 105–115). Hershey, PA: IGI Global.

Li, X. (2010). Inference degradation in information fusion: A Bayesian Network case. In Sugumaran, V. (Ed.), *Methodological advancements in intelligent information technologies: Evolutionary trends* (pp. 92–109). Hershey, PA: IGI Global.

Lian, S. (2010). Trust issues and solutions in multimedia content distribution. In Yan, Z. (Ed.), *Trust modeling and management in digital environments: From social concept to system development* (pp. 101–125). Hershey, PA: IGI Global. doi:10.4018/978-1-61520-682-7.ch005.

Liang, J., Chen, C., Zhao, H., Hu, H., & Tian, J. (2010). Gait feature fusion using factorial HMM. In Wang, L., & Geng, X. (Eds.), *Behavioral biometrics for human identification: Intelligent applications* (pp. 189–206). Hershey, PA: IGI Global.

Lilien, L., & Bhargava, B. (2009). Privacy and trust in online interactions. In Chen, K., & Fadlalla, A. (Eds.), *Online consumer protection: Theories of human relativism* (pp. 85–122). Hershey, PA: IGI Global.

Lin, K., Chen, M., Rodrigues, J. J., & Ge, H. (2012). System design and data fusion in body sensor networks. In Rodrigues, J., de la Torre Díez, I., & Sainz de Abajo, B. (Eds.), *Telemedicine and E-Health services, policies, and applications: Advancements and developments* (pp. 1–25). Hershey, PA: IGI Global. doi:10.4018/978-1-4666-0888-7.ch001.

Lin, S. (2009). A Bayesian based machine learning application to task analysis. In Wang, J. (Ed.), *Encyclopedia of data warehousing and mining* (2nd ed., pp. 133–139). Hershey, PA: IGI Global.

Lin, T. Y., Barot, R., & Tsumoto, S. (2010). Some remarks on the concept of approximations from the view of knowledge engineering. [IJCINI]. *International Journal of Cognitive Informatics and Natural Intelligence, 4*(2), 1–11. doi:10.4018/jcini.2010040101.

Lindström, J., & Hanken, C. (2012). Security challenges and selected legal aspects for wearable computing. [JITR]. *Journal of Information Technology Research, 5*(1), 68–87. doi:10.4018/jitr.2012010104.

Liu, Y. (2008). Rational concerns about biometric technology: Security and privacy. In Subramanian, R. (Ed.), *Computer security, privacy and politics: Current issues, challenges and solutions* (pp. 94–134). Hershey, PA: IGI Global. doi:10.4018/978-1-59904-804-8.ch006.

Lueth, T. C., D'Angelo, L. T., & Czabke, A. (2010). TUM-AgeTech: A new framework for pervasive medical devices. In Coronato, A., & De Pietro, G. (Eds.), *Pervasive and smart technologies for healthcare: Ubiquitous methodologies and tools* (pp. 295–321). Hershey, PA: IGI Global. doi:10.4018/978-1-61520-765-7.ch014.

Mahier, J., Pasquet, M., Rosenberger, C., & Cuozzo, F. (2009). Biometric authentication. In Khosrow-Pour, M. (Ed.), *Encyclopedia of information science and technology* (2nd ed., pp. 346–354). Hershey, PA: IGI Global.

Maiorana, E., Campisi, P., & Neri, A. (2009). Template protection and renewability for dynamic time warping based biometric signature verification. [IJDCF]. *International Journal of Digital Crime and Forensics, 1*(4), 40–57. doi:10.4018/jdcf.2009062404.

Maiorana, E., Campisi, P., & Neri, A. (2011). Cancellable biometrics for on-line signature recognition. In Li, C., & Ho, A. (Eds.), *New technologies for digital crime and forensics: Devices, applications, and software* (pp. 290–315). Hershey, PA: IGI Global. doi:10.4018/978-1-60960-515-5.ch020.

Majumder, S., & Das, T. S. (2013). Watermarking of data using biometrics. In Bhattacharyya, S., & Dutta, P. (Eds.), *Handbook of Research on computational intelligence for engineering, science, and business* (pp. 623–648). Hershey, PA: IGI Global.

Mandreoli, F., Martoglia, R., Penzo, W., Sassatelli, S., & Villani, G. (2009). Paving the way to an effective and efficient retrieval of data over semantic overlay networks. In Ma, Z., & Wang, H. (Eds.), *The semantic web for knowledge and data management* (pp. 151–175). Hershey, PA: IGI Global.

Manvi, S. S., & Kakkasageri, M. S. (2009). Emerging security issues in VANETS for e-business. In Lee, I. (Ed.), *Handbook of research on telecommunications planning and management for business* (pp. 599–614). Hershey, PA: IGI Global. doi:10.4018/978-1-60566-194-0.ch039.

Marcialis, G. L., Roli, F., Coli, P., & Delogu, G. (2010). A fingerprint forensic tool for criminal investigations. In Li, C. (Ed.), *Handbook of research on computational forensics, digital crime, and investigation: Methods and solutions* (pp. 23–52). Hershey, PA: IGI Global.

Martinez-Diaz, M., Fierrez, J., & Ortega-Garcia, J. (2010). Automatic signature verification on handheld devices. In Kurkovsky, S. (Ed.), *Multimodality in mobile computing and mobile devices: Methods for adaptable usability* (pp. 321–338). Hershey, PA: IGI Global.

Maxemchuk, N. F., Tientrakool, P., & Willke, T. L. (2009). The role of communications in cyberphysical vehicle applications. In Guo, H. (Ed.), *Automotive informatics and communicative systems: Principles in vehicular networks and data exchange* (pp. 139–161). Hershey, PA: IGI Global. doi:10.4018/978-1-60566-338-8.ch008.

Mazid, M. M., Shawkat Ali, A. B., & Tickle, K. S. (2010). Intrusion detection using machine learning: Past and present. In Ali, A., & Xiang, Y. (Eds.), *Dynamic and advanced data mining for progressing technological development: Innovations and systemic approaches* (pp. 70–107). Hershey, PA: IGI Global.

McCormack, E., Jensen, M., & Hovde, A. (2010). Evaluating the use of electronic door seals (e-seals) on shipping containers. [IJAL]. *International Journal of Applied Logistics*, *1*(4), 13–29. doi:10.4018/jal.2010100102.

Me, G., Pirro, D., & Sarrecchia, R. (2009). Strong authentication for financial services: PTDs as a compromise between security and usability. In Head, M., & Li, E. (Eds.), *Mobile and ubiquitous commerce: Advanced e-business methods* (pp. 101–114). Hershey, PA: IGI Global. doi:10.4018/978-1-60566-366-1.ch007.

Mehrotra, H., Mishra, P., & Gupta, P. (2009). Biometric identification techniques. In Khosrow-Pour, M. (Ed.), *Encyclopedia of information science and technology* (2nd ed., pp. 355–361). Hershey, PA: IGI Global.

Mey Eap, T., Hatala, M., Gaševic, D., Kaviani, N., & Spasojevic, R. (2009). Open security framework for unleashing semantic web services. In Khan, K. (Ed.), *Managing web service quality: Measuring outcomes and effectiveness* (pp. 264–285). Hershey, PA: IGI Global.

204

Mezgár, I. (2009). Building and management of trust in networked information systems. In Khosrow-Pour, M. (Ed.), *Encyclopedia of information science and technology* (2nd ed., pp. 401–409). Hershey, PA: IGI Global.

Michael, K., & Michael, M. (2009). Biometrics: In search of a foolproof solution. In Michael, K., & Michael, M. (Eds.), *Innovative automatic identification and location-based services: From bar codes to chip implants* (pp. 191–233). Hershey, PA: IGI Global. doi:10.4018/978-1-59904-795-9.ch008.

Michael, K., & Michael, M. (2009). Conclusion. In Michael, K., & Michael, M. (Eds.), *Innovative automatic identification and location-based services: From bar codes to chip implants* (pp. 485–496). Hershey, PA: IGI Global. doi:10.4018/978-1-59904-795-9.ch016.

Michael, K., & Michael, M. (2009). Historical background: From manual identification to Auto-ID. In Michael, K., & Michael, M. (Eds.), *Innovative automatic identification and location-based services: From bar codes to chip implants* (pp. 43–71). Hershey, PA: IGI Global. doi:10.4018/978-1-59904-795-9.ch003.

Michael, K., & Michael, M. (2009). Magnetic-stripe cards: The consolidating force. In Michael, K., & Michael, M. (Eds.), *Innovative automatic identification and location-based services: From bar codes to chip implants* (pp. 116–153). Hershey, PA: IGI Global. doi:10.4018/978-1-59904-795-9.ch006.

Michael, K., & Michael, M. (2009). Smart cards: The next generation. In Michael, K., & Michael, M. (Eds.), *Innovative automatic identification and location-based services: From bar codes to chip implants* (pp. 154–190). Hershey, PA: IGI Global. doi:10.4018/978-1-59904-795-9.ch007.

Mo, T., Kejun, W., Jianmin, Z., & Liying, Z. (2009). Fuzzy chaotic neural networks. In Mo, H. (Ed.), *Handbook of research on artificial immune systems and natural computing: Applying complex adaptive technologies* (pp. 520–555). Hershey, PA: IGI Global. doi:10.4018/978-1-60566-310-4.ch024.

Mogollon, M. (2008). Access authentication. In Mogollon, M. (Ed.), *Cryptography and security services: Mechanisms and applications* (pp. 152–188). Hershey, PA: IGI Global. doi:10.4018/978-1-59904-837-6.ch007.

Mohammadian, M., & Jentzsch, R. (2008). Intelligent agent framework for secure patient-doctor profiling and profile matching. [IJHISI]. *International Journal of Healthcare Information Systems and Informatics*, *3*(3), 38–57. doi:10.4018/jhisi.2008070103.

Mohammed, L. A. (2009). Security issues in pervasive computing. In Godara, V. (Ed.), *Risk assessment and management in pervasive computing: Operational, legal, ethical, and financial perspectives* (pp. 196–217). Hershey, PA: IGI Global.

Mohammed, L. A. (2011). On the design of secure atm system. In Al-Mutairi, M., & Mohammed, L. (Eds.), *Cases on ICT utilization, practice and solutions: Tools for managing day-to-day issues* (pp. 213–233). Hershey, PA: IGI Global.

Mondal, K. (2013). A novel fuzzy rule guided intelligent technique for gray image extraction and segmentation. In Bhattacharyya, S., & Dutta, P. (Eds.), *Handbook of research on computational intelligence for engineering, science, and business* (pp. 163–181). Hershey, PA: IGI Global.

Monteleone, S. (2011). Ambient intelligence: Legal challenges and possible directions for privacy protection. In Akrivopoulou, C., & Psygkas, A. (Eds.), *Personal data privacy and protection in a surveillance era: Technologies and practices* (pp. 201–221). Hershey, PA: IGI Global.

Nambiar, S., & Lu, C. (2008). M-payment solutions and m-commerce fraud management. In Ravi, V. (Ed.), *Advances in banking technology and management: Impacts of ICT and CRM* (pp. 139–158). Hershey, PA: IGI Global.

Naoe, K., Sasaki, H., & Takefuji, Y. (2012). Secure key generation for static visual watermarking by machine learning in intelligent systems and services. In Chiu, D. (Ed.), *Theoretical and analytical service-focused systems design and development* (pp. 106–121). Hershey, PA: IGI Global. doi:10.4018/978-1-4666-1767-4.ch006.

Naseer, A., & Stergioulas, L. K. (2010). Health-Grids in health informatics: A taxonomy. In Khoumbati, K., Dwivedi, Y., Srivastava, A., & Lal, B. (Eds.), *Handbook of research on advances in health informatics and electronic healthcare applications: Global adoption and impact of information communication technologies* (pp. 124–143). Hershey, PA: IGI Global.

Ng, A., Watters, P., & Chen, S. (2012). A consolidated process model for identity management. [IRMJ]. *Information Resources Management Journal*, *25*(3), 1–29. doi:10.4018/irmj.2012070101.

Okamoto, S., Sycara, K., & Scerri, P. (2009). Personal assistants for human organizations. In Dignum, V. (Ed.), *Handbook of research on multi-agent systems: Semantics and dynamics of organizational models* (pp. 514–540). Hershey, PA: IGI Global. doi:10.4018/978-1-60566-256-5.ch021.

Onwubiko, C. (2012). Challenges to managing privacy impact assessment of personally identifiable data. In Chou, T. (Ed.), *Information assurance and security technologies for risk assessment and threat management: Advances* (pp. 254–272). Hershey, PA: IGI Global.

Ouedraogo, M., Mouratidis, H., Dubois, E., & Khadraoui, D. (2011). Security assurance evaluation and it systems' context of use security criticality. [IJHCR]. *International Journal of Handheld Computing Research*, *2*(4), 59–81. doi:10.4018/jhcr.2011100104.

Pagallo, U. (2012). Responsibility, jurisdiction, and the future of "privacy by design". In Dudley, A., Braman, J., & Vincenti, G. (Eds.), *Investigating cyber law and cyber ethics: Issues, impacts and practices* (pp. 1–20). Hershey, PA: IGI Global.

Panagiotakopoulos, T., Fengou, M., Lymberopoulos, D., & Babulak, E. (2010). Ubiquitous healthcare. In Lazakidou, A. (Ed.), *Biocomputation and biomedical informatics: Case studies and applications* (pp. 254–280). Hershey, PA: IGI Global.

Pashalidis, A., & Mitchell, C. J. (2012). Privacy in identity and access management systems. In Sharman, R., Das Smith, S., & Gupta, M. (Eds.), *Digital identity and access management: Technologies and frameworks* (pp. 316–328). Hershey, PA: IGI Global.

Pasquet, M., Rosenberger, C., & Cuozzo, F. (2009). Security for electronic commerce. In Khosrow-Pour, M. (Ed.), *Encyclopedia of information science and technology* (2nd ed., pp. 3383–3391). Hershey, PA: IGI Global.

Patel, A., Taghavi, M., Júnior, J. C., Latih, R., & Zin, A. M. (2012). Safety measures for social computing in Wiki learning environment. [IJISP]. *International Journal of Information Security and Privacy*, 6(2), 1–15. doi:10.4018/jisp.2012040101.

Patlitzianas, K., & Metaxiotis, K. (2010). Formulating modern energy policy through a collaborative expert model. In Metaxiotis, K. (Ed.), *Intelligent information systems and knowledge management for energy: Applications for decision support, usage, and environmental protection* (pp. 223–241). Hershey, PA: IGI Global.

Pein, R. P., Lu, J., & Renz, W. (2013). Methods employed. In Lu, Z. (Ed.), *Design, performance, and analysis of innovative information retrieval* (pp. 301–320). Hershey, PA: IGI Global.

Pejout, N. (2010). World wide weber: Formalise, normalise, rationalise. In Maumbe, B. (Ed.), *E-Agriculture and e-government for global policy development: Implications and future directions* (pp. 267–281). Hershey, PA: IGI Global.

Peláez, J. I., Doña, J. M., & La Red, D. (2008). Fuzzy imputation method for database systems. In Galindo, J. (Ed.), *Handbook of research on fuzzy information processing in databases* (pp. 805–821). Hershey, PA: IGI Global. doi:10.4018/978-1-59904-853-6.ch033.

Penteado, R., & Boutin, E. (2008). Creating strategic information for oranizations with structured text. In do Prado, H., & Ferneda, E. (Eds.), *Emerging technologies of text mining: Techniques and applications* (pp. 34–53). Hershey, PA: IGI Global.

Peters, J. F. (2008). Monocular vision system that learns with approximation spaces. In Hassanien, A., Suraj, Z., Slezak, D., & Lingras, P. (Eds.), *Rough computing: Theories, technologies and applications* (pp. 186–203). Hershey, PA: IGI Global.

Petkovic, M., & Ibraimi, L. (2011). Privacy and security in e-Health applications. In Röcker, C., & Ziefle, M. (Eds.), *E-Health, assistive technologies and applications for assisted living: challenges and solutions* (pp. 23–48). Hershey, PA: IGI Global. doi:10.4018/978-1-60960-469-1.ch002.

Pinheiro, C., Gomide, F., Carpinteiro, O., & Lima, I. (2010). Granular synthesis of rule-based models and function approximation using rough sets. In Yao, J. (Ed.), *Novel developments in granular computing: Applications for advanced human reasoning and soft computation* (pp. 408–425). Hershey, PA: IGI Global. doi:10.4018/978-1-60566-324-1.ch017.

Pokorný, M., & Fojtík, P. (2011). Fuzzy logic based modeling in the complex system fault diagnosis. In Jozefczyk, J., & Orski, D. (Eds.), *Knowledge-based intelligent system advancements: Systemic and cybernetic approaches* (pp. 108–128). Hershey, PA: IGI Global.

Policarpio, S., & Zhang, Y. (2011). A formal language for XML authorisations based on answer set programming and temporal interval logic constraints. [IJSSE]. *International Journal of Secure Software Engineering*, 2(1), 22–39. doi:10.4018/jsse.2011010102.

Polkowski, L., & Semeniuk-Polkowska, M. (2010). On foundations and applications of the paradigm of granular rough computing. In Wang, Y. (Ed.), *Discoveries and breakthroughs in cognitive informatics and natural intelligence* (pp. 350–367). Hershey, PA: IGI Global.

Pratheepan, Y., Condell, J., & Prasad, G. (2010). Individual identification from video based on "behavioural biometrics". In Wang, L., & Geng, X. (Eds.), *Behavioral biometrics for human identification: Intelligent applications* (pp. 75–100). Hershey, PA: IGI Global.

Pugi, L., & Allotta, B. (2012). Hardware-in-the-loop testing of on-board subsystems: Some case studies and applications. In Flammini, F. (Ed.), *Railway safety, reliability, and security: Technologies and systems engineering* (pp. 249–280). Hershey, PA: IGI Global. doi:10.4018/978-1-4666-1643-1.ch011.

Rahman, H. (2009). Electronic payment systems in developing countries for improved governance system. In Rahman, H. (Ed.), *Social and political implications of data mining: Knowledge management in e-government* (pp. 126–149). Hershey, PA: IGI Global. doi:10.4018/978-1-60566-230-5.ch008.

Raisinghani, M. S., Klassen, C., & Schkade, L. L. (2009). Intelligent software agents in e-commerce. In Khosrow-Pour, M. (Ed.), *Encyclopedia of information science and technology* (2nd ed., pp. 2137–2140). Hershey, PA: IGI Global.

Raisinghani, M. S., Starr, B., Hickerson, B., Morrison, M., & Howard, M. (2008). Information technology/systems offshore outsourcing: Key risks and success factors. [JITR]. *Journal of Information Technology Research*, 1(1), 72–92. doi:10.4018/jitr.2008010107.

Rajendran, B., & Venkataraman, N. (2009). FOSS solutions for community development. [IJICTHD]. *International Journal of Information Communication Technologies and Human Development*, 1(1), 22–32. doi:10.4018/jicthd.2009010102.

Reiners, T., Wriedt, S., & Rea, A. (2011). Property-based object management and security. In Rea, A. (Ed.), *Security in virtual worlds, 3D webs, and immersive environments: Models for development, interaction, and management* (pp. 170–207). Hershey, PA: IGI Global.

Resconi, G., & Kovalerchuk, B. (2011). Agents in quantum and neural uncertainty. In Chen, S., Kambayashi, Y., & Sato, H. (Eds.), *Multi-agent applications with evolutionary computation and biologically inspired technologies: Intelligent techniques for ubiquity and optimization* (pp. 50–77). Hershey, PA: IGI Global.

Riener, A. (2012). Sitting postures and electrocardiograms: A method for continuous and non-disruptive driver authentication. In Traore, I., & Ahmed, A. (Eds.), *Continuous authentication using biometrics: Data, Models, and metrics* (pp. 137–168). Hershey, PA: IGI Global.

Ries, S. (2008). Trust and accountability. In Mühlhäuser, M., & Gurevych, I. (Eds.), *Handbook of research on ubiquitous computing technology for real time enterprises* (pp. 363–389). Hershey, PA: IGI Global. doi:10.4018/978-1-59904-832-1.ch016.

Rigatos, G. G. (2010). Technical analysis and implementation cost assessment of Sigma-Point Kalman filtering and particle filtering in autonomous navigation systems. In Rigatos, G. (Ed.), *Intelligent Industrial Systems: Modeling, automation and adaptive behavior* (pp. 125–151). Hershey, PA: IGI Global. doi:10.4018/978-1-61520-849-4.ch005.

Roberts, L. D. (2009). Cyber identity theft. In Luppicini, R., & Adell, R. (Eds.), *Handbook of research on technoethics* (pp. 542–557). Hershey, PA: IGI Global.

Roldan, M., & Rea, A. (2011). Individual privacy and security in virtual worlds. In Rea, A. (Ed.), *Security in virtual worlds, 3D webs, and immersive environments: Models for development, interaction, and management* (pp. 1–19). Hershey, PA: IGI Global.

208

Roussaki, I., Kalatzis, N., Liampotis, N., Frank, K., Sykas, E. D., & Anagnostou, M. (2012). Developing context-aware personal smart spaces. In Alencar, P., & Cowan, D. (Eds.), *Handbook of research on mobile software engineering: Design, implementation, and emergent applications* (pp. 659–676). Hershey, PA: IGI Global. doi:10.4018/978-1-61520-655-1.ch035.

Rückemann, C. (2013). Integrated information and computing systems for advanced cognition with natural sciences. In Rückemann, C. (Ed.), *Integrated information and computing systems for natural, spatial, and social sciences* (pp. 1–26). Hershey, PA: IGI Global.

Russ, M., Jones, J. G., & Jones, J. K. (2008). Knowledge-based strategies and systems: A Systematic review. In Lytras, M., Russ, M., Maier, R., & Naeve, A. (Eds.), *Knowledge management strategies: A handbook of applied technologies* (pp. 1–62). Hershey, PA: IGI Global. doi:10.4018/978-1-59904-603-7.ch001.

Russo, M. R., Bryan, V. C., & Penney, G. (2012). Emergency preparedness: Life, limb, the pursuit of safety and social justice. [IJAVET]. *International Journal of Adult Vocational Education and Technology*, *3*(2), 23–34. doi:10.4018/javet.2012040103.

Sadri, F., & Stathis, K. (2009). Ambient intelligence. In Rabuñal Dopico, J., Dorado, J., & Pazos, A. (Eds.), *Encyclopedia of artificial intelligence* (pp. 85–91). Hershey, PA: IGI Global.

Salah, A. A. (2010). Machine learning for biometrics. In Olivas, E., Guerrero, J., Martinez-Sober, M., Magdalena-Benedito, J., & Serrano López, A. (Eds.), *Handbook of research on machine learning applications and trends: Algorithms, methods, and techniques* (pp. 539–560). Hershey, PA: IGI Global.

Scholliers, J., Toivonen, S., Permala, A., & Lahtinen, T. (2012). A concept for improving the security and efficiency of multimodal supply chains. [IJAL]. *International Journal of Applied Logistics*, *3*(2), 1–13. doi:10.4018/jal.2012040101.

Sharma, K., & Singh, A. (2011). Biometric security in the e-world. In Nemati, H., & Yang, L. (Eds.), *Applied cryptography for cyber security and defense: Information encryption and cyphering* (pp. 289–337). Hershey, PA: IGI Global.

Shields, P. (2010). ICTs and border security policies in the United States and the European Union. In Adomi, E. (Ed.), *Handbook of research on information communication technology policy: Trends, issues and advancements* (pp. 373–401). Hershey, PA: IGI Global. doi:10.4018/978-1-61520-847-0.ch022.

Shields, P. (2010). ICTs and border security policies in the United States and the European Union. In Adomi, E. (Ed.), *Handbook of research on information communication technology policy: Trends, issues and advancements* (pp. 373–401). Hershey, PA: IGI Global. doi:10.4018/978-1-61520-847-0.ch022.

Shukla, A., Tiwari, R., & Rathore, C. P. (2010). Intelligent biometric system using soft computing tools. In Khosrow-Pour, M. (Ed.), *Breakthrough discoveries in information technology research: Advancing trends* (pp. 191–207). Hershey, PA: IGI Global. doi:10.4018/978-1-61692-004-3.ch014.

Singamsetty, P., & Panchumarthy, S. (2012). Automatic fuzzy parameter selection in dynamic fuzzy voter for safety critical systems. [IJFSA]. *International Journal of Fuzzy System Applications*, *2*(2), 68–90. doi:10.4018/ijfsa.2012040104.

Singh, R., Vatsa, M., & Gupta, P. (2009). Biometrics. In Pagani, M. (Ed.), *Encyclopedia of multimedia technology and networking* (2nd ed., pp. 121–127). Hershey, PA: IGI Global.

Sivagurunathan, S., Mohan, V., & Subathra, P. (2010). Distributed trust based authentication scheme in a clustered environment using threshold cryptography for vehicular ad hoc network. [IJBDCN]. *International Journal of Business Data Communications and Networking*, *6*(2), 1–18. doi:10.4018/jbdcn.2010040101.

Soar, J. (2011). Ageing, chronic disease, technology, and smart homes: An Australian perspective. In Soar, J., Swindell, R., & Tsang, P. (Eds.), *Intelligent technologies for bridging the grey digital divide* (pp. 15–29). Hershey, PA: IGI Global.

Soda, P., & Iannello, G. (2008). The relevance of computer-aided-diagnosis systems in microscopy applications to medicine and biology. In Wickramasinghe, N., & Geisler, E. (Eds.), *Encyclopedia of healthcare information systems* (pp. 1175–1182). Hershey, PA: IGI Global. doi:10.4018/978-1-59904-889-5.ch147.

Soon, C. B. (2009). Radio frequency identification history and development. In Symonds, J., Ayoade, J., & Parry, D. (Eds.), *Auto-identification and ubiquitous computing applications* (pp. 1–17). Hershey, PA: IGI Global. doi:10.4018/978-1-60566-298-5.ch001.

Stoklasa, J. (2012). A fuzzy approach to disaster modeling: Decision making support and disaster management tool for emergency medical rescue services. In Mago, V., & Bhatia, N. (Eds.), *Cross-disciplinary applications of artificial intelligence and pattern recognition: Advancing technologies* (pp. 564–582). Hershey, PA: IGI Global.

Su, S., & Li, S. (2010). Network based fusion of global and local information in time series prediction with the use of soft-computing techniques. In Wang, L., & Hong, T. (Eds.), *intelligent soft computation and evolving data mining: Integrating advanced technologies* (pp. 176–196). Hershey, PA: IGI Global. doi:10.4018/978-1-61520-757-2.ch009.

Subhasini, P., Ane, B. K., Roller, D., & Krishnaveni, M. (2012). Intelligent classifiers fusion for enhancing recognition of genes and protein pattern of hereditary diseases. In Lecca, P., Tulpan, D., & Rejaraman, K. (Eds.), *Systemic approaches in bioinformatics and computational systems biology: Recent advances* (pp. 220–248). Hershey, PA: IGI Global.

Suleiman, J., & Huston, T. (2011). Protected health information (PHI) in a small business. In Nemati, H. (Ed.), *Security and privacy assurance in advancing technologies: New developments* (pp. 106–118). Hershey, PA: IGI Global.

Sultan, Z., & Kwan, P. (2012). generalized evidential processing in multiple simultaneous threat detection in UNIX. In G. Adamson, & J. Polgar (Eds.), Enhancing enterprise and service-oriented architectures with advanced web portal technologies (pp. 104-120). Hershey, PA: IGI Global. doi: doi:10.4018/978-1-4666-0336-3.ch009.

Suomi, R., Aho, T., Björkroth, T., & Koponen, A. (2009). Biometrical identification as a challenge for legislation: The finnish case. In Godara, V. (Ed.), *Risk assessment and management in pervasive computing: Operational, legal, ethical, and financial perspectives* (pp. 233–245). Hershey, PA: IGI Global.

Swierzowicz, J. (2009). Multimedia data mining trends and challenges. In Pagani, M. (Ed.), *Encyclopedia of multimedia technology and networking* (2nd ed., pp. 965–971). Hershey, PA: IGI Global.

Tang, Z., Huang, X., & Bagchi, K. (2009). Agent-based intelligent system modeling. In Rabuñal Dopico, J., Dorado, J., & Pazos, A. (Eds.), *Encyclopedia of artificial intelligence* (pp. 51–57). Hershey, PA: IGI Global.

Tappert, C. C., Villani, M., & Cha, S. (2010). Keystroke biometric identification and authentication on long-text input. In Wang, L., & Geng, X. (Eds.), *Behavioral biometrics for human identification: intelligent applications* (pp. 342–367). Hershey, PA: IGI Global.

Tawfik, H., Huang, R., Samy, M., & Nagar, A. (2009). On the use of intelligent systems for the modelling of financial literacy parameters. [JITR]. *Journal of Information Technology Research, 2*(4), 17–35. doi:10.4018/jitr.2009062902.

Temel, T. (2011). Biologically-inspired learning: An overview and application to odor recognition. In Temel, T. (Ed.), *System and circuit design for biologically-inspired intelligent learning* (pp. 59–92). Hershey, PA: IGI Global.

Theodoridis, D. C., Boutalis, Y. S., & Christodoulou, M. A. (2013). High order neuro-fuzzy dynamic regulation of general nonlinear multi-variable systems. In Zhang, M. (Ed.), *Artificial higher order neural networks for modeling and simulation* (pp. 134–161). Hershey, PA: IGI Global.

Thomas, C., & Balakrishnan, N. (2012). Usefulness of sensor fusion for security incident analysis. In Onwubiko, C., & Owens, T. (Eds.), *Situational awareness in computer network defense: Principles, methods and applications* (pp. 165–180). Hershey, PA: IGI Global. doi:10.4018/978-1-4666-0104-8.ch010.

Thuraisingham, B., Tsybulnik, N., & Alam, A. (2009). Administering the semantic web: Confidentiality, privacy and trust management. In Nemati, H. (Ed.), *Techniques and applications for advanced information privacy and security: Emerging organizational, ethical, and human issues* (pp. 262–277). Hershey, PA: IGI Global. doi:10.4018/978-1-60566-210-7.ch017.

Tian, Y., Feris, R., Brown, L., Vaquero, D., Zhai, Y., & Hampapur, A. (2010). Multi-scale people detection and motion analysis for video surveillance. In Wang, L., Cheng, L., & Zhao, G. (Eds.), *Machine learning for human motion analysis: Theory and practice* (pp. 107–132). Hershey, PA: IGI Global.

Tivive, F. H., & Bouzerdoum, A. (2008). A brain-inspired visual pattern recognition architecture and its applications. In Verma, B., & Blumenstein, M. (Eds.), *Pattern recognition technologies and applications: Recent advances* (pp. 244–264). Hershey, PA: IGI Global. doi:10.4018/978-1-59904-807-9.ch011.

Tiwari, S., Singh, A., Singh, R. S., & Singh, S. K. (2012). Internet security using biometrics. In Prakash Vidyarthi, D. (Ed.), *Technologies and protocols for the future of internet design: Reinventing the Web* (pp. 114–142). Hershey, PA: IGI Global. doi:10.4018/978-1-4666-0203-8.ch006.

Tiwari, S., & Singh, S. K. (2013). Information security governance using biometrics. In Mellado, D., Enrique Sánchez, L., Fernández-Medina, E., & Piattini, M. (Eds.), *IT security governance innovations: Theory and research* (pp. 191–224). Hershey, PA: IGI Global.

Trajkovski, G., Stojanov, G., Collins, S., Eidelman, V., Harman, C., & Vincenti, G. (2009). Cognitive robotics and multiagency in a fuzzy modeling framework. [IJATS]. *International Journal of Agent Technologies and Systems, 1*(1), 50–73. doi:10.4018/jats.2009010104.

Traoré, I., & Ahmed, A. A. (2012). Introduction to continuous authentication. In Traore, I., & Ahmed, A. (Eds.), *Continuous authentication using biometrics: Data, models, and metrics* (pp. 1–22). Hershey, PA: IGI Global.

Travieso González, C. M., & Morales Moreno, A. (2009). Non-cooperative facial biometric identification systems. In Rabuñal Dopico, J., Dorado, J., & Pazos, A. (Eds.), *Encyclopedia of artificial intelligence* (pp. 1259–1265). Hershey, PA: IGI Global.

Trcek, D. (2009). E-business systems security in intelligent organizations. In Khosrow-Pour, M. (Ed.), *Encyclopedia of information science and technology* (2nd ed., pp. 1222–1226). Hershey, PA: IGI Global.

Tsatsoulis, P. D., Jaech, A., Batie, R., & Savvides, M. (2012). Multimodal biometric hand-off for robust unobtrusive continuous biometric authentication. In Traore, I., & Ahmed, A. (Eds.), *Continuous authentication using biometrics: Data, models, and metrics* (pp. 68–88). Hershey, PA: IGI Global.

Tyagi, S., Sirohi, P., Khan, M. Y., & Darwish, A. (2012). industrial information security, safety, and trust. In M. Khan, & A. Ansari (Eds.), Handbook of research on industrial informatics and manufacturing intelligence: Innovations and solutions (pp. 20-31). Hershey, PA: IGI Global. doi: doi:10.4018/978-1-4666-0294-6.ch002.

Vildjiounaite, E., Rantakokko, T., Alahuhta, P., Ahonen, P., Wright, D., & Friedwwald, M. (2008). Privacy threats in emerging ubicomp applications: Analysis and safeguarding. In Mostefaoui, S., Maamar, Z., & Giaglis, G. (Eds.), *Advances in ubiquitous computing: Future paradigms and directions* (pp. 316–347). Hershey, PA: IGI Global. doi:10.4018/978-1-59904-840-6.ch012.

Vincenti, G., & Braman, J. (2009). Hybrid emotionally aware mediated multiagency. In Trajkovski, G., & Collins, S. (Eds.), *Handbook of research on agent-based societies: Social and cultural interactions* (pp. 199–214). Hershey, PA: IGI Global. doi:10.4018/978-1-60566-236-7.ch014.

Vincenti, G., & Trajkovski, G. (2009). Fuzzy mediation in shared control and online learning. In Król, D., & Nguyen, N. (Eds.), *Intelligence integration in distributed knowledge management* (pp. 263–285). Hershey, PA: IGI Global.

Vouyioukas, D., & Maglogiannis, I. (2010). Communication issues in pervasive healthcare systems and applications. In Coronato, A., & De Pietro, G. (Eds.), *Pervasive and smart technologies for healthcare: Ubiquitous methodologies and tools* (pp. 197–227). Hershey, PA: IGI Global. doi:10.4018/978-1-61520-765-7.ch010.

Walkerdine, J., Phillips, P., & Lock, S. (2009). A tool supported methodology for developing secure Mobile P2P systems. In B. Seet (Ed.), Mobile peer-to-peer computing for next generation distributed environments: Advancing conceptual and algorithmic applications (pp. 283-300). Hershey, PA: IGI Global. doi: doi:10.4018/978-1-60566-715-7.ch013.

Wang, C., Chang, R., & Ho, J. (2011). Collaborative video surveillance for distributed visual data mining of potential risk and crime detection. In Koyuncugil, A., & Ozgulbas, N. (Eds.), *Surveillance technologies and early warning systems: Data mining applications for risk detection* (pp. 194–204). Hershey, PA: IGI Global. doi:10.4018/978-1-61350-101-6.ch313.

Wang, H. (2009). Survivability evaluation modeling techniques and measures. In Gupta, J., & Sharma, S. (Eds.), *Handbook of research on information security and assurance* (pp. 504–517). Hershey, PA: IGI Global.

Wang, S., Chang, S., & Wang, R. (2011). Applying the linguistic strategy-oriented aggregation approach to determine the supplier performance with ordinal and cardinal data forms. [IJFSA]. *International Journal of Fuzzy System Applications, 1*(2), 1–16. doi:10.4018/ijfsa.2011040101.

Wang, Y. (2009). Statistical opportunities, roles, and challenges in network security. In Wang, Y. (Ed.), *Statistical techniques for network security: Modern statistically-based intrusion detection and protection* (pp. 1–34). Hershey, PA: IGI Global.

Wang, Y., Berwick, R. C., Haykin, S., Pedrycz, W., Kinsner, W., & Baciu, G. et al. (2013). Cognitive informatics and cognitive computing in year 10 and beyond. In Wang, Y. (Ed.), *Cognitive informatics for revealing human cognition: Knowledge manipulations in natural intelligence* (pp. 140–157). Hershey, PA: IGI Global.

Wang, Y., Bhatti, M. W., & Guan, L. (2010). Neural networks for language independent emotion recognition in speech. In Wang, Y. (Ed.), *Discoveries and breakthroughs in cognitive informatics and natural intelligence* (pp. 461–484). Hershey, PA: IGI Global.

Wautelet, Y., Schinckus, C., & Kolp, M. (2010). A modern epistemological reading of agent orientation. In Sugumaran, V. (Ed.), *Methodological advancements in intelligent information technologies: Evolutionary trends* (pp. 43–55). Hershey, PA: IGI Global.

Webb, K. (2009). Security implications for management from the onset of information terrorism. In Knapp, K. (Ed.), *Cyber security and global information assurance: Threat analysis and response solutions* (pp. 97–117). Hershey, PA: IGI Global. doi:10.4018/978-1-60566-326-5.ch005.

Weiss, G. (2009). Data mining in the telecommunications industry. In Wang, J. (Ed.), *Encyclopedia of data warehousing and mining* (2nd ed., pp. 486–491). Hershey, PA: IGI Global.

West, R., Mayhorn, C., Hardee, J., & Mendel, J. (2009). The weakest link: A psychological perspective on why users make poor security decisions. In Gupta, M., & Sharman, R. (Eds.), *Social and human elements of information security: Emerging trends and countermeasures* (pp. 43–60). Hershey, PA: IGI Global.

Willow, C. C. (2009). Next-generation strategic business model for the U.S. internet service providers: Rate-based internet subscription. [IJITN]. *International Journal of Interdisciplinary Telecommunications and Networking*, *1*(3), 31–41. doi:10.4018/jitn.2009070102.

Wong, W. K., Loo, C. K., & Lim, W. S. (2010). Quaternion based machine condition monitoring system. In Rigatos, G. (Ed.), *Intelligent industrial systems: Modeling, automation and adaptive behavior* (pp. 476–508). Hershey, PA: IGI Global. doi:10.4018/978-1-61520-849-4.ch017.

Wu, Q., McGinnity, M., Prasad, G., & Bell, D. (2009). Knowledge discovery in databases with diversity of data types. In Wang, J. (Ed.), *Encyclopedia of data warehousing and mining* (2nd ed., pp. 1117–1123). Hershey, PA: IGI Global.

Xiang, Y., & Tian, D. (2010). Multi-core supported deep packet inspection. In Li, K., Hsu, C., Yang, L., Dongarra, J., & Zima, H. (Eds.), *Handbook of research on scalable computing technologies* (pp. 858–873). Hershey, PA: IGI Global.

Xie, C., Turnquist, M. A., & Waller, S. T. (2012). A hybrid lagrangian relaxation and tabu search method for interdependent-choice network design problems. In Montoya-Torres, J., Juan, A., Huaccho Huatuco, L., Faulin, J., & Rodriguez-Verjan, G. (Eds.), *Hybrid algorithms for service, computing and manufacturing systems: Routing and scheduling solutions* (pp. 294–324). Hershey, PA: IGI Global.

Xu, G., Ma, Y., & Zhang, Z. (2010). An efficient and automatic iris recognition system using ICM neural network. In Wang, Y. (Ed.), *Discoveries and breakthroughs in cognitive informatics and natural intelligence* (pp. 445–460). Hershey, PA: IGI Global.

Yadav, N., & Poellabauer, C. (2012). Challenges of mobile health applications in developing countries. In Watfa, M. (Ed.), *E-Healthcare systems and wireless communications: Current and future challenges* (pp. 1–22). Hershey, PA: IGI Global.

Yadav, S. B. (2008). SEACON: An integrated approach to the analysis and design of secure enterprise architecture-based computer networks. [IJISP]. *International Journal of Information Security and Privacy*, 2(1), 1–25. doi:10.4018/jisp.2008010101.

Yager, R. R. (2011). On possibilistic and probabilistic information fusion. [IJFSA]. *International Journal of Fuzzy System Applications*, 1(3), 1–14. doi:10.4018/ijfsa.2011070101.

Yamamoto, G. T. (2010). Problems. In Yamamoto, G. (Ed.), *Mobilized marketing and the consumer: Technological developments and challenges* (pp. 198–219). Hershey, PA: IGI Global.

Yampolskiy, R. V., & Govindaraju, V. (2010). Taxonomy of behavioural biometrics. In Wang, L., & Geng, X. (Eds.), *Behavioral biometrics for human identification: Intelligent applications* (pp. 1–43). Hershey, PA: IGI Global.

Yan, G., Olariu, S., & Salleh, S. (2010). A probabilistic routing protocol in VANET. [IJMCMC]. *International Journal of Mobile Computing and Multimedia Communications*, 2(4), 21–37. doi:10.4018/jmcmc.2010100102.

Yang, S. Q., & Xu, A. (2012). Applying semantic web technologies to meet the relevant challenge of customer relationship management for the U.S. academic libraries in the 21st century using 121 e-agent framework. In Colomo-Palacios, R., Varajão, J., & Soto-Acosta, P. (Eds.), *Customer relationship management and the social and semantic web: Enabling cliens conexus* (pp. 284–311). Hershey, PA: IGI Global.

Yayilgan, S. Y., Blobel, B., Petersen, F., Hovstø, A., Pharow, P., Waaler, D., & Hijazi, Y. (2012). An architectural approach to building ambient intelligent travel companions. [IJEHMC]. *International Journal of E-Health and Medical Communications*, 3(3), 86–95. doi:10.4018/jehmc.2012070107.

Yu, S., & Wang, L. (2010). Gait recognition and analysis. In Wang, L., & Geng, X. (Eds.), *Behavioral biometrics for human identification: intelligent applications* (pp. 151–168). Hershey, PA: IGI Global.

Zarri, G. P. (2012). A conceptual methodology for dealing with terrorism "narratives". In Li, C., & Ho, A. (Eds.), *Crime prevention technologies and applications for advancing criminal investigation* (pp. 274–290). Hershey, PA: IGI Global. doi:10.4018/978-1-4666-1758-2.ch018.

Zeng, C., Jia, W., He, X., & Xu, M. (2013). Recent advances on graph-based image segmentation techniques. In Bai, X., Cheng, J., & Hancock, E. (Eds.), *Graph-based methods in computer vision: Developments and applications* (pp. 140–154). Hershey, PA: IGI Global.

214

Zhang, D., Song, F., Xu, Y., & Liang, Z. (2009). Decision level fusion. In Zhang, D., Song, F., Xu, Y., & Liang, Z. (Eds.), *Advanced pattern recognition technologies with applications to biometrics* (pp. 328–348). Hershey, PA: IGI Global. doi:10.4018/978-1-60566-200-8.ch015.

Zhang, D., Song, F., Xu, Y., & Liang, Z. (2009). Feature level fusion. In Zhang, D., Song, F., Xu, Y., & Liang, Z. (Eds.), *Advanced pattern recognition technologies with applications to biometrics* (pp. 273–304). Hershey, PA: IGI Global. doi:10.4018/978-1-60566-200-8.ch013.

Zhang, D., Song, F., Xu, Y., & Liang, Z. (2009). From single biometrics to multi-biometrics. In Zhang, D., Song, F., Xu, Y., & Liang, Z. (Eds.), *Advanced pattern recognition technologies with applications to biometrics* (pp. 254–272). Hershey, PA: IGI Global. doi:10.4018/978-1-60566-200-8.ch012.

Zhang, D., Song, F., Xu, Y., & Liang, Z. (2009). Matching score level fusion. In Zhang, D., Song, F., Xu, Y., & Liang, Z. (Eds.), *Advanced pattern recognition technologies with applications to biometrics* (pp. 305–327). Hershey, PA: IGI Global. doi:10.4018/978-1-60566-200-8.ch014.

Zhao, X., Zuo, M. J., & Moghaddass, R. (2013). Generating indicators for diagnosis of fault levels by integrating information from two or more sensors. In Kadry, S. (Ed.), *Diagnostics and prognostics of engineering systems: Methods and techniques* (pp. 74–97). Hershey, PA: IGI Global. doi:10.4018/978-1-4666-2770-3.ch015.

Zhao, Y., & Yao, Y. (2010). User-centered interactive data mining. In Wang, Y. (Ed.), *discoveries and breakthroughs in cognitive informatics and natural intelligence* (pp. 110–125). Hershey, PA: IGI Global.

Zhou, Z., Wang, H., & Lou, P. (2010). Sensor integration and data fusion theory. In Zhou, Z., Wang, H., & Lou, P. (Eds.), *Manufacturing intelligence for industrial engineering: Methods for system self-organization, learning, and adaptation* (pp. 160–188). Hershey, PA: IGI Global. doi:10.4018/978-1-60566-864-2.ch007.

About the Authors

Marina L. Gavrilova is an Associate Head and Professor in the Department of Computer Science, University of Calgary. Dr. Gavrilova's research interests lie in the area of computational geometry, image processing, optimization, and spatial and biometric modeling. Prof. Gavrilova is founder and co-director of two innovative research laboratories: the Biometric Technologies Laboratory: Modeling and Simulation; and the SPARCS Laboratory for Spatial Analysis in Computational Sciences. Prof. Gavrilova's publication list includes over 120 journal and conference papers, edited special issues, books and book chapters, including World Scientific Bestseller of the Month (2007), *Image Pattern Recognition: Synthesis and Analysis in Biometric*, and Springer book, *Computational Intelligence: A Geometry-Based Approach*. Together with Dr. Kenneth Tan, Prof. Gavrilova founded ICCSA series of successful international events in 2001. She founded and chaired International Workshop on Computational Geometry and Applications since 2000, was Co-Chair of the International Workshop on Biometric Technologies BT 2004, Calgary; served as Overall Chair of the 3rd International Conference on Voronoi Diagrams in Science and Engineering (ISVD) in 2006, was Organization Chair of WADS 2009 (Banff), and General Chair of International Conference on Cyberworlds CW2011 (October 4-6, Banff, Canada). Prof. Gavrilova is an Editor-in-Chief of successful *LNCS Transactions on Computational Science Journal*, Springer-Verlag since 2007 and serves on the Editorial Board for the *International Journal of Computational Sciences and Engineering*, *CAD/CAM Journal*, and *Journal of Biometrics*. She has received numerous awards and was successful in obtaining major funding for her research program. Her research was profiled in newspaper and TV interviews, most recently being chosen together with five outstanding Canadian scientists to be featured in National Museum of Civilization and National Film Canada production. Her greatest accomplishment, in her own words, is finding a delicate balance between a professional and personal life while striving to give her best to both. Together with her husband, Dr. Dmitri Gavrilov, she is a proud parent of two wonderful boys, Andrei and Artemy.

Maruf Monwar is a Post-Doctoral Fellow at the Department of Electrical and Computer Engineering, Carnegie Mellon University, USA. He received his PhD in Computer Science from the University of Calgary, Canada, B.Sc. (Hons.) and M.Sc. in Computer Science & Technology at the University of Rajshahi, Bangladesh, and M.Sc. in Computer Science at the University of Northern BC, Canada. He is an Assistant Professor at the Department of Computer Science & Engineering, University of Rajshahi, Bangladesh. His primary research interests include pattern matching, biometric fusion, expression recognition, and biological data processing. He is one of the recipients of the prestigious inaugural NSERC Vanier CGS Scholarship and NSERC Post-Doctoral Fellowship. He has served as a guest editor for the International Journal of Biometrics.

Index

3D model 151

A

artificial entity 113-114, 119-120, 122, 124, 131, 140, 143, 166
Artificial Intelligence (AI) 2
Artificial Neural Networks (ANN) 133
artimetric 122
Automated Turing Test (ATT) 2
avatar behavior 116

B

behavioral characteristic 3, 11, 113, 124
behavioral identifier 10
betweenness centrality 157
biometric authentication system 7, 14, 21, 56, 80-81, 96
Borda count method 51, 69-70, 73-76, 78, 80, 86, 89, 106

C

Chaotic Neural Network (CNN) 1, 3-5, 14, 23, 44, 130, 132-135, 139, 142-143, 145-147, 165
Chaotic Simulated Annealing method (CSA) 135
circumvention 3, 12, 21, 131-132, 142
classifier and decision fusion 55-56
closeness centrality 157-158
collectability 12, 48
computational intelligence 1, 3-4, 62, 111, 145, 164-165
computational intelligence paradigm 4, 165
Condorcet Criterion 67, 79, 86-89, 98
Condorcet Looser 86
Copeland Method 88-89
Cumulative Match Characteristics (CMC) 18
cyberinfrastructure 115
cybersecurity 19, 113, 120, 124

D

data acquisition module 13
Decision Level Fusion 57, 61, 63, 107-108, 164
decision module 13-15, 70
defuzzification 99, 101
degree centrality 157
Delaunay Triangulation (DT) 36, 140
Dempster combination rule 62, 101
Deoxyribonucleic Acid (DNA) 9
dimensionality reduction 59, 130-132, 136-137, 143-144, 147, 150, 155
distinctiveness 12, 48, 136

E

eigenfaces 28, 35, 44-45, 128, 137-139
eigenimage 26, 29
Eigenvector centrality 157-158
enzyme activity 83
Equal Error Rate (EER) 18-19, 110
evolutionary computing 1-2, 15

F

facial database 57, 92
Facial Recognition Technology (FERET) 92
Failure-to-Capture Rate (FCR) 18
Failure-to-Enroll Rate (FER) 18
False Accept Rate (FAR) 18
False Reject Rate (FRR) 18
feature extraction module 13
Feature Level Fusion 57, 59, 66-67, 164
fisherimage 26, 29, 103
full rank aggregation 71
fusion module 4, 31, 63, 69, 101-103
fuzzification 99, 101
Fuzzy fusion 48, 61-64, 78, 98-102, 104-110, 165
fuzzy fusion-based biometric system 98
fuzzy fusion mechanism 98, 110